Game Development with Blender and Godot

Leverage the combined power of Blender and Godot for building a point-and-click adventure game

Kumsal Obuz

BIRMINGHAM—MUMBAI

Game Development with Blender and Godot

Group Product Manager: Rohit Rajkumar

Publishing Product Manager: Nitin Nainani

Senior Editor: Hayden Edwards

Senior Content Development Editor: Rashi Dubey

Technical Editor: Joseph Aloocaran

Copy Editor: Safis Editing

Project Coordinator: Sonam Pandey

Proofreader: Safis Editing

Indexer: Tejal Daruwale Soni

Production Designer: Nilesh Mohite

Marketing Coordinator: Teny Thomas

First published: September 2022

Production reference: 1080922

Published by Packt Publishing Ltd.

Livery Place

35 Livery Street

Birmingham

B3 2PB, UK.

ISBN 978-1-80181-602-1

www.packt.com

To my wife, Becky, for helping me during times when I didn't even think I needed it. To my parents, Hulya and Ilhan, for providing the opportunities that played a big part in who I am.

– Kumsal Obuz

Contributors

About the author

Kumsal Obuz is a self-taught veteran web developer with more than 15 years of experience in two different countries, leading teams and projects of various sizes.

After several years of preparation, he started his own game studio, Viroid Games, in August 2020. He then launched a small puzzle-strategy game at the end of 2020 and is currently working on an ambitious farming simulation game.

He also enjoys mentoring, which runs in the family since both of his parents are teachers. In 2019, he founded and still organizes the Godot Toronto group on Discord.

In his spare time, he likes reading history (mostly medieval) and science-fiction.

I want to thank my friend and mentor, Gokhan Ercan, for generously sharing his wisdom with me. Also, big thanks to the lovely people at Shmooz for their pleasant snacks, coffee, and most importantly, their camaraderie.

About the reviewers

Anthony Cardinale is a software engineer who specialized in 3D and video game development for over 10 years. He has worked for large groups, notably on the development of 3D virtual reality experiences.

In parallel to his various missions, Anthony is an entrepreneur and shares his knowledge through online video courses or books.

He has written many books in French on 3D modeling with Blender and video game development with Unity and Godot.

Joseph B. Manley

Table of Contents

Part 2: Asset Management

6

7

8

Part 3: Clara's Fortune – An Adventure Game

9

Designing the Level   149

10

Making Things Look Better with Lights and Shadows   179

11

Creating the User Interface   203

12

Interacting with the World through Camera and Character
Controllers   225

13

Finishing with Sound and Animation   257

14

Conclusion   291

Preface

Game Development with Godot and Blender is a comprehensive introduction for those who are new to building 3D models and games, allowing you to leverage the abilities of these two technologies to create dynamic, interactive, and engaging games.

This book will start by focusing on what low-poly modeling actually is, before diving into using Blender to create, rig, and animate our models. We will also make sure that these assets are game-ready, making it easy for you to import them into Godot and use your assets effectively and efficiently. Then, in Godot, you will use the game engine to design scenes, work with light and shadows, and transform your 3D models into interactive, controllable assets.

By the end of the book, you will have a seamless workflow between Blender and Godot that is specifically geared towards game development and will have created a point-and-click adventure game following our instructions and guidance. Beyond this point, you should be able to take these newly acquired skills and create your own 3D games from conception to completion!

Who this book is for

This book is for game developers who are looking to make the transition from 2D to 3D games. You should have a basic understanding of Godot, and be able to navigate the UI, understand the Inspector panel, create scenes, add scripts to game objects, and so on. Previous experience with Blender is helpful but not required.

What this book covers

Chapter 1, Creating Low-Poly Models, covers the creation of low-poly models in Blender. You'll also look at how to utilize modifiers to expedite the process.

Chapter 2, Building Materials and Shaders, shows you how to create and assign different materials to your models, and understand where shaders come into play.

Chapter 3, Adding and Creating Textures, teaches you how to prepare your models for texturing. Applying third-party textures and creating your own are also covered in this chapter.

Chapter 4, Adjusting Cameras and Lights, presents different light types and how to capture a shot of your scene. You'll be revisiting some of these notions in the Godot context later in *Chapter 10, Making Things Look Better with Lights and Shadows*.

Chapter 5, Setting up Animation and Rigging, discusses the notion of animation and whether doing it in Godot or Blender is the right choice. Once we settle the matter in Blender's favor, you'll rig and animate a simple model.

Chapter 6, Exporting Blender Assets, tackles a most crucial and often ignored topic: exporting your models from Blender. You'll be specifically shown a format that is the most suitable for Godot Engine.

Chapter 7, Importing Blender Assets into Godot, conveniently shows how to import your models into Godot. The transition between different applications is not always smooth, so you'll also be presented with shortcomings and workarounds.

Chapter 8, Adding Sound Assets, investigates the use of sound in Godot Engine. You'll partake in a short exercise to play a sound file after discovering different types of audio files the engine supports.

Chapter 9, Designing the Level, will be the beginning of a series of exercises for building a point-and-click adventure game. To kick off the effort, you'll be designing the level with the models that come within the GitHub repository.

Chapter 10, Making Things Look Better with Lights and Shadows, presents different light types you can deploy in your level to enhance the look and feel of the game. To complement the scene further, you'll also discover the use of global illumination and post-processing effects.

Chapter 11, Creating the User Interface, discusses the necessity of user interfaces. Then, you'll utilize a bunch of Godot UI components to compose a piece of note. Last but not least, you'll investigate why creating themes in Godot might be a time-saver.

Chapter 12, Interacting with the World through Camera and Character Controllers, presents different camera types and settings on different gaming platforms. After attaining a basic view into the game world, you'll continue with detecting user input, which is essential for the type of game you are building. To finish off, you'll use this information to move a game character to their designated spot.

Chapter 13, Finishing with Sound and Animation, finishes the core mechanics of our little game. To that end, you'll be adding sound effects and animations to certain game objects. Also, you'll create a simple animation in Godot and create the necessary conditions for the player to meet in order to trigger this animation. Once all the in-game requirements are finished, you'll load a new level for the player.

Chapter 14, Conclusion, shows how to export your game to Windows, so you can share it with the world. You'll finish this chapter and the book off by getting to know what else Godot can offer to you.

To get the most out of this book

You will need the Windows versions of Blender 2.93 and Godot 3.4.4 installed on your computer. All the visual examples and code samples have been tested for these versions. If you have newer or older versions installed, you might notice discrepancies.

Software/hardware covered in the book	Operating system requirements
Blender 2.93	Windows
Godot 3.4.4	

Knowing how to use GitHub at a basic level might help. Alternatively, you can download the whole repository and work with your local copy.

If you are using the digital version of this book, we advise you to type the code yourself or access the code from the book's GitHub repository (a link is available in the next section). Doing so will help you avoid any potential errors related to the copying and pasting of code.

Download the example code files

You can download the example code files for this book from GitHub at `https://github.com/PacktPublishing/Game-Development-with-Blender-and-Godot`. If there's an update to the code, it will be updated in the GitHub repository.

We also have other code bundles from our rich catalog of books and videos available at `https://github.com/PacktPublishing/`. Check them out!

Download the color images

We also provide a PDF file that has color images of the screenshots and diagrams used in this book. You can download it here: `https://packt.link/0KyZi`.

Conventions used

There are a number of text conventions used throughout this book.

`Code in text`: Indicates code words in text, database table names, folder names, filenames, file extensions, pathnames, dummy URLs, user input, and Twitter handles. Here is an example: "If you increase the radius to `10.0`, something interesting will happen."

A block of code is set as follows:

```
extends AudioStreamPlayer

func _unhandled_key_input(event: InputEventKey) -> void:
    if event.is_pressed() and event.scancode == KEY_SPACE:
        stream_paused = false
    else:
        stream_paused = true
```

Bold: Indicates a new term, an important word, or words that you see onscreen. For instance, words in menus or dialog boxes appear in **bold**. Here is an example: "When you applied the **Solidify** modifier, you must have seen that there are so many other modifiers."

> **Tips or important notes**
> Appear like this.

Get in touch

Feedback from our readers is always welcome.

General feedback: If you have questions about any aspect of this book, email us at customercare@ packtpub.com and mention the book title in the subject of your message.

Errata: Although we have taken every care to ensure the accuracy of our content, mistakes do happen. If you have found a mistake in this book, we would be grateful if you would report this to us. Please visit www.packtpub.com/support/errata and fill in the form.

Piracy: If you come across any illegal copies of our works in any form on the internet, we would be grateful if you would provide us with the location address or website name. Please contact us at copyright@packt.com with a link to the material.

If you are interested in becoming an author: If there is a topic that you have expertise in and you are interested in either writing or contributing to a book, please visit authors.packtpub.com.

Part 1: 3D Assets with Blender

This part of the book provides you a detailed look into how to create models, textures, and animation in Blender. By the end of this part, you'll be able to create game-ready assets.

In this part, we cover the following chapters:

- *Chapter 1, Creating Low-Poly Models*
- *Chapter 2, Building Materials and Shaders*
- *Chapter 3, Adding and Creating Textures*
- *Chapter 4, Adjusting Cameras and Lights*
- *Chapter 5, Setting Up Animation and Rigging*

1
Creating Low-Poly Models

Blender is a sophisticated program that has gone through a lot of iterations to get to the point where it is now. More and more professionals in different industries are investigating it as an alternative to other well-known 3D applications out there, such as Maya, 3ds Max, ZBrush, and Modo. Also, Blender happens to be a good starting point for hobbyists and people who can't afford the licensing fees of the aforementioned software. Additionally, Blender has a helpful and large community that creates courses and tutorials. **Blender Conference (BCON)** is an annual event where you can meet professionals.

An important decision you must make before you start creating 3D content with any type of software is where you are going to use your assets – this directly affects the style and workflow you will follow to accomplish the task. One type of workflow is called low-poly modeling, with which you create 3D assets that have a minimum number of details.

In this chapter, we'll discuss why low-poly modeling might be beneficial compared to other workflows. Following the advantages, you'll learn how to create low-poly assets using different techniques. We'll conclude this chapter by introducing a few modifiers that might prove indispensable.

In this chapter, we will cover the following main topics:

- Understanding low-poly models
- Advantages of low-poly models
- Creating a low-poly barrel
- Automating with modifiers

Technical requirements

To follow the instructions in the chapters that involve Blender, you must install the necessary software on your computer. The Blender website – more specifically, their download page at `https://www.blender.org/download/` – contains instructions and links for your platform. In this book, we are using Blender 2.93. Although version 3.0 will offer interesting and exciting options to new and existing Blender users, the current version is more than capable of creating game assets for your projects and the topics covered in this book.

This book uses GitHub to store the code that will be used in the Godot chapters. However, the same repository (`https://github.com/PacktPublishing/Game-Development-with-Blender-and-Godot`) also hosts the Blender files used throughout the relevant chapters. Where it makes sense, the repository will be structured with `Start` and `Finish` folders inside each specific chapter for you to start over or compare your work as you make progress.

Understanding low-poly models

Simply put, a 3D model is considered to be low-poly when it uses the minimum number of polygons to make its most characteristic features, mainly its look and feel. However, let's take a look at them in a little more detail.

In real-time applications such as game engines, your computer's **central processing unit (CPU)** and **graphics processing unit (GPU)** are responsible for processing the visual information you see on the screen. In the last two decades, the trend has been leaning heavily toward the GPU side since GPUs are dedicated to one main task: processing graphics.

GPUs have an advantage over CPUs in that regard, and they don't discriminate between 2D and 3D graphics. However, whereas 2D images contain pixel information, 3D objects are represented by data that holds the necessary coordinate information that defines the object.

Although a cube is still a bunch of pixels after it is rendered on your screen, the data that defines the cube is essentially eight points, which are called **vertices**. For demonstration purposes, in the following screenshot, Blender's vertex size setting has been changed so that you can see where those vertices are more easily:

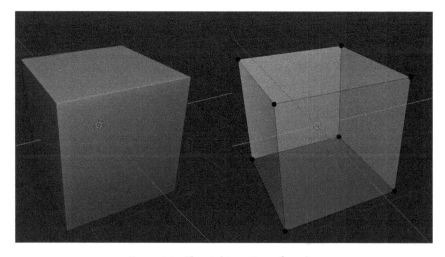

Figure 1.1 – The eight vertices of a cube

Both cubes are the same object, but it's possible to render the same eight vertices and their relationship with each other in two different ways: one that looks like a solid object (on the left) and another that looks transparent (on the right). So, keep in mind that vertices are data points that define the shape of the object, not how it looks. Later in this chapter, you'll learn how to make objects look different, similar to what's shown in the preceding screenshot.

Before we discuss what makes a model low-poly, we must understand a few other essential parts. You've already seen that the vertex is the most crucial component, but there are two more concepts you must be aware of:

- Edge
- Face

Let's see how these two relate to a vertex. By doing so, we'll be on our way to understanding what makes a model low-poly.

Parts of a 3D model

An **edge** is what connects two vertices. It's as simple as that. If you look at *Figure 1.1* again, you'll see that not all the vertices are connected. However, when they are connected, it's called an edge and depicted by Blender with a straight line going from one vertex to the other.

A **face**, as you may have deduced, is a logical outcome when you arrange vertices – or edges – in a certain pattern. For example, a cube or a six-sided die has six faces. Although *Figure 1.1* makes it look like you need four edges to make a face, there is a simpler way – that is, three edges are enough to form a face. So, a triangle is the simplest face, also known as a **polygon**.

Low or hi, what's the difference?

When you are designing a model, you are creating polygons. The density of the polygons in a model is what determines whether a model can be considered low-poly. The following figure displays one low-poly and one high-poly work sample, courtesy of Sketchfab users *MohammadRezae* and *DJS_05*:

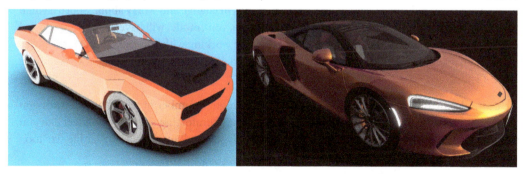

Figure 1.2 – An example of a low-poly versus high-poly model

You can find a lot of examples of different polygon counts on websites such as Sketchfab.

In the industry, if you are asking for a model to be designed for you, you may want to mention that you want it done in low-poly form. It's often agreed that if you don't mention this, people will assume it's going to have as many polygons as possible since you would want your models to be as detailed as possible with plenty of polygons. So, the distinction is made when you are cutting those polygons out, not when you are keeping them in.

Let's focus on our default cube again. Is it low-poly or hi-poly? It might be both. Although we know that only eight vertices are needed to create a cube, we could have had many more vertices along the edges that connected the original corner vertices. However, it would not have made any difference in the rendered result. That being said, it would have taken the computer a lot longer to render the same visual result.

So, as mentioned previously, when your model has just enough polygons to make sense of the object you'd like to design, you'll have a low-poly model.

Although GPUs are fast and they do a fantastic job these days of rendering millions of polygons and going low-poly may feel like you are cutting corners, there are good reasons why you may not want to have that many polygons in your project.

Advantages of low-poly models

Here is a quick list of the benefits of following a low-poly modeling practice:

- Fewer polygons
- Small file size
- A certain artistic style
- Easy to prototype
- No or minimal texturing

Working with fewer polygons certainly means fewer things to change and worry about. Shortly, you'll learn how to create a barrel, and by the end of that exercise, your model will have close to a thousand polygons. This number may seem high at first but imagine working with a hi-poly barrel model with more than 10,000 polygons. So, if you are new to 3D modeling, low-poly modeling is a great place to start.

Should you decide to alter your models, working with a higher number of polygons will force you to be more careful. So, in essence, having fewer polygons is comforting since you will feel like you have more control over your creation. Naturally, fewer polygons will result in a smaller file size too.

The artistic style advantage is a non-technical item in the advantages list. Nevertheless, it might be an important decision. Let's focus on *Figure 1.3*, for example. You'll see why lack of detail doesn't always mean lack of imagination:

Figure 1.3 – Low-poly model landscape

Here, you can see just enough details to figure out that there is a church. Perhaps this church is looking onto a town square. The mountain tops have some snow. Is this a peaceful town that's appealing to tourists for winter sports? Perhaps the townspeople are currently hiding in the church from a villain? Our imagination fills in the details. Whatever the case and the game genre is, the low-poly aspect of the 3D models doesn't induce a penalty for creativity. In fact, in the last few years, we've seen more games with low-poly assets making headlines.

If you are working in a small game development team or if you are the only developer, you'll sometimes want to focus on game mechanics first to see if the idea is fun. In situations like these, you'll want to prototype objects quickly so that you can embed them into your code. When the model you are working on has a generic shape of the object you would like to design and has enough details, then you might be done. That's why it's a highly sought-after choice among indie developers since you can move forward quickly to the next model, then to programming your game. In essence, low-poly modeling is like prototyping but it's a few steps more refined than placing a cone for a tree, a cylinder for a barrel, or a cube for a crate.

Last on the list is **texturing**. This is a process where you give a certain look and feel to your model. A sandy beach usually looks yellow. If it's a rocky beach, then the rocks will most likely have different tones of gray. Thus, it's about mainly applying color information to the surfaces of your model. Sometimes, this color information will be complemented by additional data such as reflectivity, metallicity, and roughness. We'll discover all this in the next chapter.

It's often said that most things in the computer world are a trade-off. Speed versus quality versus price is a common example where you can most likely have two out of three but not all three. Despite all the benefits a low-poly workflow provides, there are some limitations, but recognizing them will help you to find workarounds or plan ahead.

Limitations of low-poly models

If your models need to show damage such as missing parts along an edge or some chunks blown out of a face, then you need to introduce more polygons in those areas. This still won't make it a high-poly model, but you've got to consider additional polygons if you fancy some dynamic details.

Also, if you decide to animate your low-poly models, you'll need to introduce more geometry by adding more polygons in the areas where there will be bending and twisting (depending on the model you are animating).

Additionally, since there are fewer polygons, you may have to be creative with the lighting of your scene to give the illusion of detail. Although the color of the water in *Figure 1.3* is the same throughout the composition, the designer used a couple of clever methods to make the scene look more interesting. First, the water's surface looks fractured. This gives the illusion that there is some slight movement in this water's body. Perhaps there is a gentle breeze. Second, some of those fractures have a reflective material applied. This makes the surface reflect the objects further ahead on the horizon.

We'll look at ways to overcome these limitations in the following chapters, but for now, let's learn how to create a few low-poly models of our own.

Creating a low-poly barrel

Every discipline comes with a few conventions for beginners. If you are learning a new programming language, writing "Hello World" to the screen is a classic example. Learning how to use 3D modeling software is no different. For example, a barrel, a potion bottle, or a donut can be started with basic shapes you are familiar with, such as a cylinder, a cone, or a torus, respectively.

In this section, you'll learn how to design a barrel but, first, here are a few useful shortcuts that will help you navigate around and accomplish the tasks we'll cover in this section:

- **Rotate**: Middle mouse button + drag mouse
- **Zoom**: Scroll mouse wheel forward/backward
- **Move**: *Shift + Drag Mouse*

Blender is rich with so many shortcuts and it's possible to change them to your liking once you gain more experience. Speaking of shortcuts, this book only lists Windows shortcuts. However, when you see the *Ctrl* key mentioned, it's the *Command* key in macOS.

When you launch Blender for the first time, you'll be presented with some options. One important option is to decide which mouse button to use to select objects. Historically, the right mouse button was the default, but you may find this uncommon. If you dismissed that initial screen, and you are not happy with the mouse button assignment for the select operation, you can still change it by going to the **Edit** menu and selecting **Blender Preferences**. In the **Keymap** section, expand the **Preferences** section, as shown in the following screenshot; you'll be able to change a bunch of settings, including **Select with Mouse Button**:

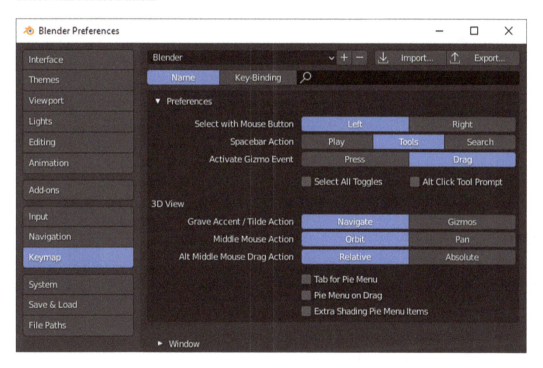

Figure 1.4 – The Preferences window of Blender

Speaking of the select button, whichever side you choose, the other side will be reserved for moving the **3D cursor** to a new position. A 3D cursor is a visual marker you place in the world. When you add new elements to your scene at a particular location, or things need to align to a certain point, the 3D cursor will be that point. We'll most likely keep the 3D cursor where it is for most exercises, but keep in mind that if the left click is for selection, then the right click is for the 3D cursor, and vice versa.

> **Official manual**
>
> Since this book is about game development, we'll focus on a small and relevant portion of Blender. However, sometimes, looking at the official manual might be a good idea, especially for shortcuts. The Blender website has a decent user manual: `https://docs.blender.org/manual/en/2.93/`.

Modeling is a multi-step process. It involves starting with the basics and adding more details as you go. The following is what we'll do to design a barrel:

- Start with a primitive
- Edit the model
- Shape the body
- Separate the lid
- Finish the body
- Place metal rings
- Finalize the lid

The list is merely an example workflow that highlights useful parts of Blender. When you gain more experience and find a different order to accomplish what you have in mind, you can work in whatever way works for you. However, you are likely to start with primitives.

Starting with a primitive

A new scene in Blender comes with a cube, a camera, and a light source. Since we are going to create a barrel that is more like a cylinder, we should get rid of that cube:

1. Select the cube and press *X* on your keyboard to delete it.
2. Trigger the **Add** menu to the left of the **Object** menu.
3. Select **Cylinder** under the **Mesh** group.

The shortcut for adding new objects is *Shift + A*, which will bring up the same list of options. If you feel like deleting the other default objects, feel free to do so since you can always add them later using the **Add** menu. The following screenshot shows where you can find it:

Figure 1.5 – You can add many types of primitives to your scene

Once you add the cylinder to your scene, you'll see that the cylinder comes with a lot of side faces; 32 to be exact. For a low-poly barrel, that's a lot of faces that could be cut down by half and you would still have a decent-looking barrel.

When you add a new object, a panel will appear at the bottom left of the screen. The title of this panel will reflect what you are currently trying to accomplish. In this case, it should display **Add Cylinder**. If it looks closed, click the title and it'll expand to show the properties you can alter for your cylinder.

The default options are all fine except for the number of vertices. However, this is also a good chance to play with the values and see the changes reflect instantly. While you are doing all this, that panel may disappear if you click away from your cylinder. To bring it back, click **Adjust Last Operation** under the **Edit** menu. When you feel like you've got the hang of editing a new object's properties, you can set the relevant values, as shown in the following screenshot:

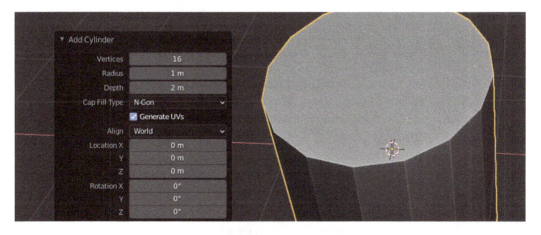

Figure 1.6 – 16 faces will be enough for creating our barrel

Adding a primitive such as a cylinder has introduced a new object to your scene. You've changed its basic properties, such as the number of vertices. That number defines how many points are used to make up the top and bottom circles, as shown in the preceding screenshot. This was all done at the object level; hence, you've been working in **Object Mode**. Now, it's time to dive deeper and edit the finer details of this cylinder.

Editing the model

It may seem like every time you change the value of something, you are editing the model. However, from Blender's perspective, not all edits are the same. When you start with primitives, there are higher-level operations you can perform such as changing the number of vertices that define the general shape of the primitive. This is what you have seen and done so far – you've been editing objects.

When you want to have more control over the vertices, faces, and edges that make up the object, you should switch to another mode that allows you to work with these properties so that you can have much more refined control over the shape of the model.

> **Mac shortcuts**
>
> You can always use menus, buttons, and other interface elements to do your work, but you'll eventually depend on shortcuts. If the shortcuts that have been mentioned so far don't work for you, then you may want to check out Blender's manual to find the right combination for your platform: `https://docs.blender.org/manual/en/2.93/interface/keymap/introduction.html`.

Select the barrel and press *Tab*. This will turn on **Edit Mode**. If you keep pressing the *Tab* key, you'll go back and forth between **Object Mode** and **Edit Mode**. You'll also see that Blender's UI is either hiding some of the buttons and menus or revealing some new ones, depending on which mode is active. This means some options are only available in a certain mode. If you are wondering where that thing you just saw disappeared, make sure you are in the right mode.

Then, in **Edit Mode**, press *Ctrl + R* to trigger **Loop Cut and Slide**. This is a context-sensitive operation, so if you see nothing happening, it's because the mouse is not over a face for this tool to operate. Hover your mouse over different parts of the cylinder. You'll see a yellow line going all the way around; the direction of the line depends on where your cursor is on that face. While still over one of the side faces, trigger your mouse wheel up twice to increase the number of cuts to 3. This is a preview of the loop cuts, but they are not part of the cylinder yet.

A loop cut will require two mouse clicks, regardless of how many loops you would like to have. With the *first click*, you are telling Blender that you want to introduce some cuts; in this case, 3. The *second click* will finalize the position of these cuts, but you can change it by moving your mouse up and down along the side of the barrel. So, in between the first and the second click, you have some freedom to position the cuts. The following screenshot shows what we are after:

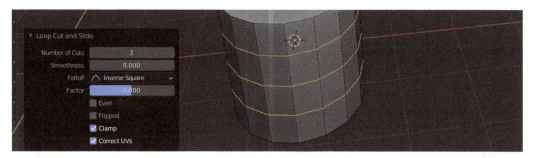

Figure 1.7 – Adding more edges with precise values

If you accidentally moved your cursor in between two clicks, which would have moved the baseline of the cuts, do not worry. Once the edges have been added, the operation's details will be displayed so that you can fine-tune where the cuts appear in your model. The important part is to set **Factor** to **0** so that you have the perfect cut in the middle. If you made a last moment change before you made the cuts, you can also adjust the number of cuts.

The main reason why you switched to **Edit Mode** is to have more control over the shape of your objects. While still in **Edit Mode**, you'll now learn how to use those loop cuts to give your object the shape of a barrel.

Shaping the body

A barrel is such a generic concept. However, we have not discussed what kind of barrel we will be working on. Technically, we are not too far off from an oil barrel since they usually look cylindrical and have two rounded-off ridges. Then, there are plastic barrels that you see in gardens for collecting rain. These tend to have a plain side with the top and bottom slightly tapered in or with the middle section slightly bulging out, depending on which way you look at it.

We'll go for a more classic one: a wooden barrel. Since we have the basic shape, we can now start adding more details to our barrel. Two things come to mind easily. Most barrels have a few metal rings – in the middle, near the bottom, and at the top – for enduring the stress of what they are holding. Also, the lid is rarely flush with the side but more likely inset, so maybe we should treat that top part separately. Let's start tackling all these one at a time.

> **Are your 3D objects looking flat?**
>
> It'd be nice to have some life in all that gray! If the default look for 3D objects feels too flat and you'd rather see the edges emphasized like you see them in pictures, here is a trick. There is a button with a down-looking icon at the top-right corner of the **3D Viewport**. If you click that button and expand the **Viewport Shading** panel, you can switch **Lighting** to **MatCap**, and turn on both the **Shadow** and **Cavity** options in the panel. Selecting **Both** for the **Cavity** type may also be a good option. Investigate different values as you see fit so that you have an easier time working with your models.

Our barrel needs a belly. We need to make those loops we have just introduced wider to create a classic shape for the barrel. With those three edges still selected, hit *S*, type 1 . 1, and press *Enter* to scale it up by 10%. As usual, the last operation's fine-tuning settings will be shown if you would like to adjust your values after finishing the action. Now, we only need to make the middle ring slightly larger.

Although we have been in **Edit Mode** so far, we have not investigated what you can edit. In the top-left corner of **3D Viewport** next to the **Edit Mode** dropdown, you'll see *Vertex*, *Edge*, and *Face* icons from left to right. These buttons have *1*, *2*, and *3* as shortcuts, respectively.

Switch to **Edge edit mode** by pressing the middle icon or *2*. To create the belly for the barrel, you need to select and scale up all the edges that make up the middle ring, but you probably don't want to do that for each edge one by one. Thus, we need to look at how to select an **edge loop**.

There are two ways to select an edge loop. The first method uses a keyboard shortcut:

1. Hold the *Alt* key.
2. Click one of the edges.

This should select all the edges that are connected to the one you've just clicked, as shown in the following screenshot:

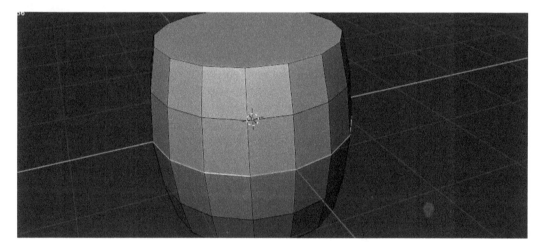

Figure 1.8 – Selecting all the edges that make a loop is easy

The second way is as follows:

1. Select one edge.
2. Go to the **Select** menu.
3. Expand **Select Loops** and choose **Edge Loops**.

Whichever way you do this, after you select the middle edge loop, you must do the following:

1. Scale it by pressing *S*.

2. Type `1.05`.

3. Hit *Enter*.

This should result in a classic barrel shape.

However, the top face still belongs to the cylinder. Although conceptually, a lid might be considered an essential part of a barrel, from an editing perspective, it must be treated as a separate object. Let's learn how to separate parts to edit them individually.

Separating the lid

To create the lid, first, make sure you are still in **Edit Mode**. Switch to face select mode by clicking the third icon next to the **Edit Mode** dropdown or by pressing *3*. Then, do the following:

1. Select the top face.

2. Press *P*.

3. Choose **Selection**.

This will separate the top face and make it a separate object.

Alternatively, you can expand the **Separate** group under the **Mesh** menu. The following screenshot shows where you can find this option if you are doing the separation with the menus:

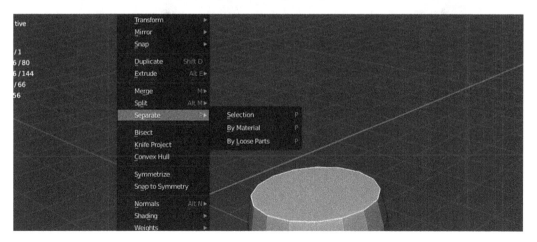

Figure 1.9 – Separating things is sometimes necessary and, in fact, helpful

Shortcuts

At this point, you must have noticed that Blender uses a lot of shortcuts. It might be difficult to learn and remember all these at the beginning. If you have a rough idea about what you'd like to do with the meshes, vertices, edges, and such, you should check out the appropriate menus near the top to see what operations are available. Pressing a shortcut key will show you just the relevant part of those menus, but investigating those menus and looking at the shortcut might be a good exercise.

For example, the *P* key is used to separate things, but there are three types of separation, so you'll still have to make a final decision on the type. However, using the shortcut still takes a shorter time than expanding the menus.

Now might be a good time to introduce you to **Outliner** in the top-right corner. The following screenshot shows all the objects that exist in your scene right now:

Figure 1.10 – The lid and the body should be two separate objects

You can ignore the **Camera** and **Light** objects if you kept them in your scene since we'll discover what those two do later in this book. Over time, when you create more objects, you'll want to label your objects so that you can easily find them in **Outliner**.

Let's try it now. Double-click the label for **Cylinder** in **Outliner** and type Body. Do the same thing for **Cylinder.001** and mark it as *Lid*. You'll also notice that clicking labels in **Outliner** will select the objects in **3D Viewport** and vice versa. Finally, hit that eye icon to hide the lid for now. We'll finalize the lid once we deal with the body.

Finishing the body

What would you say is wrong or missing from the body? It looks paper-thin, doesn't it? If only there was a way to stretch each face out or in, and fill in the gaps so it looks solid! So far, you've been selecting edges and faces. You can follow a similar workflow to select some faces, duplicate them, and move them around to give thickness to the body. This is tempting, but let's find an easy way to solidify the body.

For this, you need to enable the **Modifiers** panel. A modifier is a tool that offers a non-destructive way to change your objects. You'll get to read about a few of them in the *Automating with modifiers* section.

There is a wrench icon on the right-hand side near **3D Viewport** that is going to let you add modifiers. Here are the steps you must take to give substance to the barrel's body:

1. Switch to **Object Mode**.
2. Select the **Body** object.
3. Open the **Modifiers** panel.
4. Choose **Solidify** from the **Add Modifier** dropdown.

Modifiers change objects, so even if you are in **Edit Mode**, working with a modifier will look as if you are in **Object Mode** for the object you are editing. You'll discover some of the modifiers in that dropdown list later in this chapter. For now, the following screenshot shows what the **Solidify** modifier is doing. Most things in Blender come with a lot of values to tweak, but you only need to change the **Thickness** value in the **Solidify** options for the time being:

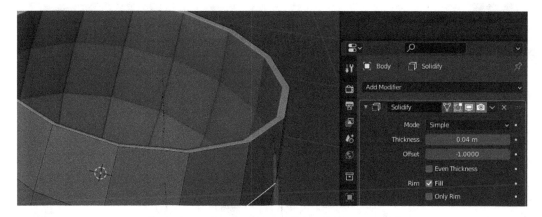

Figure 1.11 – Our barrel is starting to look more solid

How much **Thickness** is enough? 0.03 m or 0.04 m might be a good value. You could pick an industry-standard thickness or choose a value that looks visually appealing. Depending on the type of game you are working on or whether you are creating assets for a client, you can pick what works best for the asset.

A discussion about units

Most of the world is using the metric system these days. However, if either because it's the default option or a matter of preference, you may have **Imperial** units set up in your Blender copy. Throughout this book, the **Metric** system will be utilized. You can find **Units** as a panel inside the fifth tab from the top on the right-hand side. This tab contains an icon with a cone, a sphere, and what looks like a dot.

Modifiers are very helpful, but you need to get your hands dirty sometimes. This means that there is a limit to what modifiers can do for you. For example, we now need to put metal rings around the body. There is no modifier to do this for you. Nevertheless, we can still take advantage of modifiers as we go. But, first, let's create some metal rings.

Placing metal rings

The barrel now has some substance, but it's missing metal rings. Creating another cylinder and sizing it up so that it looks like a ring is too much work and requires precision. There is a simpler method that takes advantage of the barrel's geometry. You'll be using familiar methods you've already seen: loop cuts, loop selection, and separation.

While in **Edit Mode**, create a loop cut between the bottom and the first edge loop of the body. For the other loop cut, you'll be creating the cut in between the top edges and the loop right below it. In the end, you'll be creating two loop cuts, as shown in the following screenshot

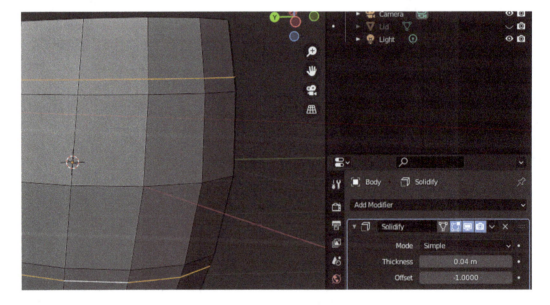

Figure 1.12 – Two separate cuts from both ends approaching the middle section

You've already seen how to select an edge loop: it involved holding the *Alt* key and clicking an edge. You'll do something very similar except it'll be for selecting a face loop. For this, make sure the face icon is clicked in **Edit Mode**. Alternatively, while in **Edit Mode**, you can press *3*.

When you hold the *Alt* key and click an edge, you'll be selecting the faces that are adjacent to that edge you've just clicked. It'll also keep selecting the other faces that are in a similar direction to complete a loop. Try it a few times with horizontal and vertical edges to see how the loops' direction changes accordingly.

What you must do is select all the faces that make up the two rings that are close to the top and bottom of the barrel. The following screenshot shows which faces should be selected so that you can separate them to form the metal rings:

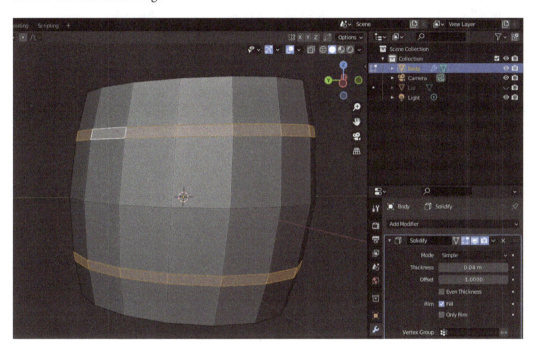

Figure 1.13 – You've got to have something selected so that you can separate it

Once you select the first loop, you can hold down *Shift* and repeat the previous operation to keep adding more loops to your selection.

Now, you are ready to separate those faces. Hit *P* to bring up the **Separate** options and choose **Selection**. Now, you can rename the newly created **Ring** object. If you go back to **Object Mode**, you'll see that you can select each object individually. Select the ring; you'll see that the **Solidify** modifier still exists for this new object too. Isn't that handy?

The thickness value in the modifier is the same, but what would happen if we changed the sign of that value? If you click the **Thickness** field, it will let you type in a value. Adjust it so it shows - 0 . 0 4 m. As you can see, it's still the same thickness, but in the other direction – it looks like we have those metal rings around the body of our barrel finally!

Now, let's learn how to add another ring for the middle section. You can follow similar steps to create two more loops, one above and one below the center loop. However, you can do better.

Select the middle edges by conducting an edge loop operation and then triggering the **Offset Edge Slide** option under the **Edge** menu or pressing *Shift + Ctrl + R*. This is very similar to **Loop Cut and Slide** but it has two major differences. First, this operation will consider an edge as its baseline and move the new edges off in opposite directions. Second, you need to click just once when you are happy with where the new edges will sit. Choosing **0.1** for the **Factor** value in the operation's properties might be a good number if you're having sensitivity issues with your mouse.

We'll follow a similar procedure: select and separate. In face edit mode, you will use a combination of *Alt + Shift* by clicking one of the vertical edges sandwiched between your new loops. After you separate the middle faces, you'll be left with an important decision: should you rename your new object and invert the direction of thickness in its modifier just like you did for the upper and lower rings? In essence, you want your new object to join its fellows. That's exactly what you'll do next but with a clever trick without repeating yourself.

> **Which mode?**
>
> During the modeling process, there are times when you'll need to edit parts of your model. In this case, being in **Edit Mode** will be necessary. However, when you separate chunks from your models, you'll most likely want to go back to **Object Mode** to do something with this new object. So, going back and forth between these two modes will be necessary and feel natural after a while.

In **Object Mode**, first, you must select the middle ring you have just created. You don't need to rename it; you'll see why shortly. You must add one more object to your selection by holding *Shift* and clicking the ring in **3D Viewport**. Make sure your last click is on the ring object you created a while ago. The order of clicks matters at this point. The last object you interact with will be considered as the active object by Blender. It will be marked with a yellow outline compared to orange outlined objects, which are part of the selection but not considered active objects.

Once you have your rings selected in the correct order, you must join them by pressing *Ctrl + J*. Did you notice what just happened? Let's break it down:

- You can no longer see **Body.001** in the **Outliner**
- The **Ring** object has accepted **Body.001** into the fellowship
- The **Ring** object's **Solidify** modifier has been applied to **Body.001**

There are no longer separate pieces since all those separate parts are now considered as one object, as shown in the following screenshot:

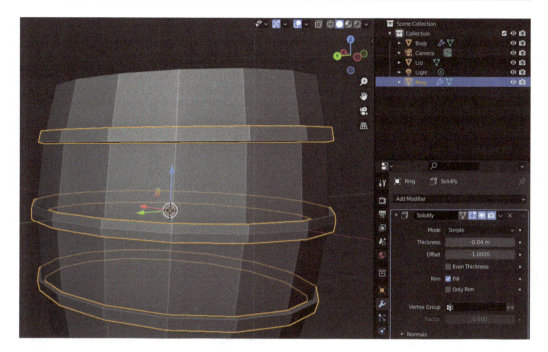

Figure 1.14 – One ring to join them all

As you get more confident in Blender, you'll find that you can follow different methods to achieve the same result. There is no right or wrong way, but rather time-saving habits, and you'll develop your own since every designer has their preferred way of doing things. Sometimes, there are other concerns, technical or artistic, that will limit your workflow. However, as a beginner, you should observe how other artists are creating similar objects. Luckily, there are plenty of examples out there, so learn, experiment, and divert as you go.

Earlier, you had to separate the lid. After that, you made changes to the body and even added rings. Now, it's time to put a lid on your barrel.

Finalizing the lid

If you hid the lid once you separated it, you can click the eye icon in **Outliner** to turn it on. You need to do a basic scale operation to put the lid in its place. To achieve this, first, select the lid, and then do the following:

1. Press *S*.
2. Type in `0.96`.
3. Press *Enter*.

Why such a precise value? Because we've been using 0.04 m in the **Solidify** modifier. So, we should reduce the scale of the lid by 4%. This will save us from the trouble of lining up all the edges of the lid so that they are flush with the inner side of the barrel. If you have been using a different value in your modifier, you've got to compensate your scale value in this step so that both add up to 1 in the end.

You've done it! With the lid in the right place and looking just below the rim level, the barrel is complete. Check out the following screenshot and compare it to your creation:

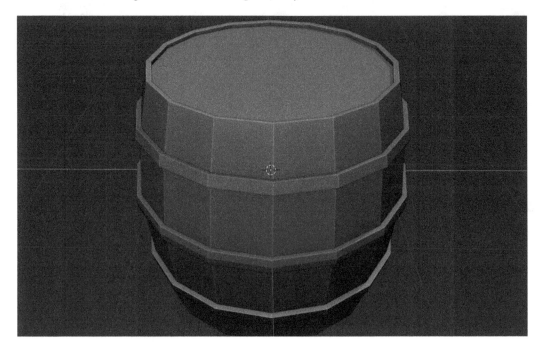

Figure 1.15 – A wooden – rather gray – barrel in its glory

If you decide to create this barrel from scratch again, perhaps you can place the upper ring close to the lid, and the lower ring at the bottom. Placing five rings is also a possibility, but you may want to adjust the height of each ring in case the composition looks busy.

So far, you have used one modifier, and it has served you well. Let's dive into more modifiers and see how powerful they can be.

Automating with modifiers

A modifier is a non-destructive way of applying an operation to change an object's geometry. This is often preferred when you don't want to take repetitive steps or the operation is complex enough that you don't want to directly alter the object's geometry.

When you applied the **Solidify** modifier, you must have seen that there are so many other modifiers. Could you imagine what you can do with each one? How about you use a few modifiers in a row? Yes, you read that right. You can stack up many modifiers and create complex shapes with little effort.

However, there is an important detail you must pay attention to – their order matters! New modifiers are always added at the bottom, and they work in conjunction with the previous modifiers in the stack. Thus, the effect is compounding. If you logically stack your modifiers, you could create something as complex as what's shown in the following screenshot with only a few primitive objects in no time:

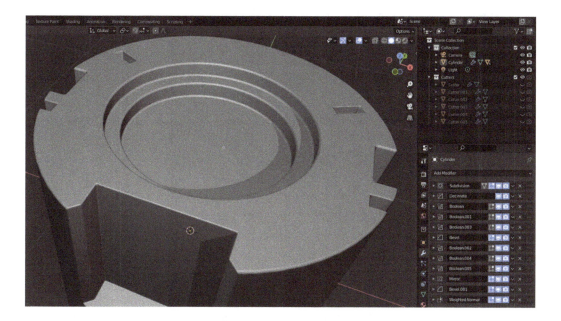

Figure 1.16 – Modifiers help you create something this complex easily

This object is using primitives such as cylinders and cubes, but the result looks interesting. This is thanks to a hefty list of modifiers and the order that they've been used. Some of the modifiers have been applied multiple times with different values, but here is a list:

- **Subdivision**
- **Decimate**
- **Boolean**
- **Bevel**
- **Mirror**
- **Weighted Normal**

At the time of writing, Blender has over 50 modifiers. Describing each would fill a book. Most likely, you'll stick with the modifiers that are in the **Generate** category. Here is a set of modifiers you'll use most of the time:

- **Boolean**: This is one of those modifiers that is used a lot and it comes in three sub-modes:

  - **Difference**: Subtracts the value of one object from another

  - **Union**: Will combine both objects

  - **Intersect**: Will only keep what's common in both meshes

- **Bevel**: Sometimes, you want to have more detail, especially along sharp edges so that they don't look too harsh – the more surface there is for the light sources to reflect on, the more realistic it'll look to the eye. This modifier will also work on vertices if you want to soften those sharp corners.

- **Array**: This makes copies of the object it's assigned to in different axes, with or without some offset if you wish. You could have a fixed number of copies or fill a particular length with as many copies as you can fit in that distance.

- **Mirror**: This is like the **Array** modifier except it creates one copy along the axis you select. You can pick multiple axes. Thus, it's possible to start with only a quarter of the object and mirror it on the X and Y axis so that you have one whole object. This allows you to keep your changes to a minimum in the original quarter so that you can mirror your changes to the rest of the mesh.

When you add your modifiers, it's sometimes not obvious which order you should stack them in. Luckily, it's possible to change their order or temporarily disable them by using the buttons that are part of the modifier's header.

While creating rings for the barrel, you could have used a different technique to achieve the same result: **extrusion**. This would require you to select what needs to be extruded – in this case, all the faces that make up the ring – and extrude along each face's outward-facing direction. Extrusion, in essence, is a technical term for moving vertices, faces, or edges.

Modifiers have a big advantage compared to classic methods such as pushing and pulling vertices and faces around. Wouldn't it be convenient to come back later and fine-tune your changes further? If you happen to select the lid now and come back to the **Body** object, the modifier will still be there. You won't have this kind of flexibility with permanent mesh modifying techniques such as extrusion.

Summary

In this chapter, you learned about the benefits of low-poly modeling. Then, you created a wooden barrel from a primitive cylinder and incorporated modifiers. Although textures may give a more realistic look to your models, you also know you can do without them.

As an exercise, feel free to create a potion bottle. You can start with a cylinder, just like you did for the barrel. The loop cuts and the scaling down values will be different to give it a conical shape. This is your chance to practice modifiers. A finished potion bottle is waiting for you in this book's GitHub repository if you want to see a finished example and compare yours.

Several shortcuts are commonly used by many professionals during the modeling process. Here is a list you've used so far:

- *Shift + A*: Add an object
- *Tab*: Switch between **Edit Mode** and **Object Mode**
- *Ctrl + R*: Introduce loop cuts
- *Ctrl + J*: Join
- *S*: Scale
- *P*: Separate

In the next chapter, you'll learn how to apply materials to your models so that parts of your model can still have a different look and feel without textures.

Further reading

The section's title suggests reading sources, but sometimes seeing is even better. Just as a picture is worth a thousand words, a video might be worth a thousand pictures. So, here is a list of URLs for video content that might be useful for all levels of Blender practitioners:

- `https://www.youtube.com/c/JoshGambrell`
- `https://www.youtube.com/c/CurtisHolt`
- `https://www.youtube.com/c/GrantAbbitt`
- `https://www.youtube.com/c/SouthernShotty`

2
Building Materials and Shaders

According to Wikipedia, a material is a substance, or a mixture of substances, that constitutes an object. This definition for real-life objects also stands true for the models you create electronically with some extra technical details. Let's look at the definition of a material in our context.

In Blender, **materials** are essentially containers that hold a bunch of numbers, colors, and textures, besides other useful stuff, and most importantly the **shader** itself. A shader is a piece of code that tells the rendering engine, either Blender's or Godot Engine's, what to do with a material's properties.

In essence, a material is like a box full of little items, and it comes with a user manual (the shader) so that the software you work with knows what to do with those little items.

You now know the raw definition of what materials and shaders are, but what are they used for? The barrel you created in the previous chapter had *metal rings* and *wooden slats* that gave it its form. However, everything in that model looked rather gray. Adding materials to your models will enhance their form by showing colors and other properties you are familiar with from real life.

In this chapter, you'll learn how to make your models look more real by applying materials. To that end, we will cover the following topics:

- Introducing materials
- Creating materials
- Assigning materials
- Discovering shaders

By the end of this chapter, you'll know how to create and assign different materials, and understand where shaders come into play during this process.

Technical requirements

Although this chapter is about materials, you'll need at least one 3D model. This can be your finished work from *Chapter 1*, *Creating Low-Poly Models*. Alternatively, you can use the barrel model that comes in the `Start` folder of the `Chapter 2` folder in this book's GitHub repository: `https://github.com/PacktPublishing/Game-Development-with-Blender-and-Godot`.

Introducing materials

As we mentioned in the introduction, materials are assigned to objects. However, you can't assign materials to all objects. When you start up a new Blender file, it comes with a cube, a camera, and a light object. Only one of these objects has substance from Blender's perspective, and that's the cube. Let's break this down a bit more to understand why it matters. Although a camera and a light source have physical properties and they occupy space in real life, this isn't the case in Blender. They are conceptual objects.

A camera is a tool through which you see the world. So, you don't get to see the visual properties of the camera itself. It doesn't matter if the camera is painted red or blue. Similarly, a light source shines a bright or dim light, sometimes with a certain color, but it doesn't take up space in a Blender scene. Therefore, if there is no substance, we can't apply a material to these two objects.

If only there was an easier way to know which objects can receive materials…

If you select each of the default objects one after another, you'll see that some icons are popping in and out of the view on the right-hand side of the screen. Different sets of options, represented by icons, are stacked in the **Properties** panel. This panel will display the relevant properties of the selected objects.

When you select the cube, you'll notice that the **Properties** panel introduces a lot of icons, different than the ones for a camera or a light object. Either hover over the icons to see their title or simply click the icons to take a quick look at what's at your disposal. While doing that, you'll eventually discover the second-last icon, which should turn on the **Materials** panel (if you need to remember it later, it's the icon that looks like a sphere with a checkerboard pattern).

You haven't created a material yet. However, Blender starts with a default cube, which comes with a default material. Let's learn how to change its color. After selecting the cube, follow these steps:

1. Open the **Materials** tab in the **Properties** panel.
2. Click the colored rectangle on the right-hand side of **Base Color**.
3. Pick a different color from the color wheel.

The following screenshot will help you find all this. Once you have chosen a color, the change will not apply to the cube at first; you'll find out why soon:

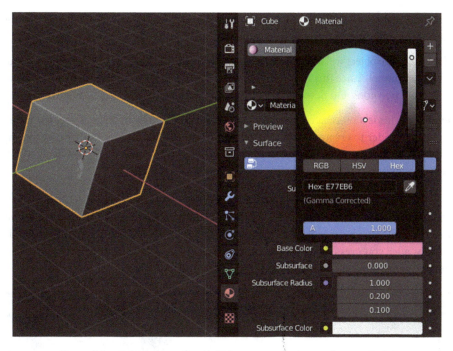

Figure 2.1 – Color is one of many things you can change for a material

While selecting a color, the color wheel will help you out. However, if you want to be more precise with the color you are selecting, the three buttons (**RGB**, **HSV**, and **Hex**) under the color wheel can help you. In the preceding screenshot, a value of **E77EB6** was used in **Hex** mode. All these color modes act like units, but the result will always be the same color when you switch between different modes.

> **Panels and settings**
>
> Working in Blender doesn't always mean you have to directly modify the geometry (vertices, edges, and faces) of your models; you will often find yourself looking for settings and altering them in many panels. Later in this book, when you work with objects such as camera and light, you'll use the appropriate panels so that you can tweak the settings for these objects.

Let's figure out why the last change you made didn't reflect on the screen. By default, Blender shows models as solid objects. Sometimes, just like an X-ray may help a doctor understand what's going on, you will need to see your model differently. The following are four different ways you can see your objects:

- **Solid**: The default option; you've been using this all along. It simply shows your model as a solid object.

- **Material Preview**: You will mainly see the color you applied to the object, but you will also see some of the other properties you have applied.

- **Wireframe**: The object will look like a metal wire has been bent and welded to create a frame that defines the model. Since this mode only renders edges and vertices, it is useful when you want to visualize polygons and detect overloaded areas so that you can easily optimize the models.

- **Rendered**: This is a more accurate view than **Material Preview** because it uses the rendering engine of Blender to create the most accurate representation. It does this by considering the lights and shadows in your scene. Naturally, it uses more GPU, so you'll most likely work with other view options most of the time.

The previous list shows all the options you have for **Viewport Shading**. The default view, **Solid**, is fast but not accurate when you want to work with materials. Now that you've changed the properties of your material, you must be in **Material Preview** to see it in effect. To switch to it, press *Z* and then *2*. Alternatively, after you press *Z*, you can select the appropriate option with your mouse, as shown in the following screenshot:

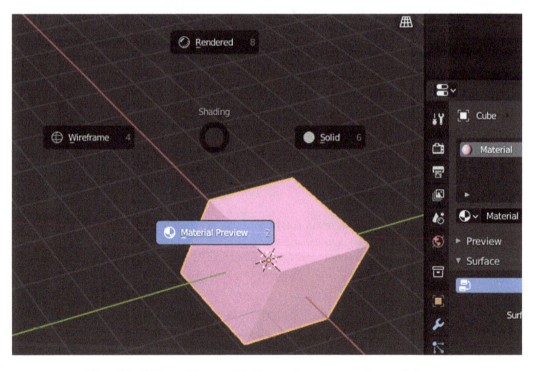

Figure 2.2 – Different Viewport Shading options presented in a radial menu

Now that you have a basic understanding of what materials are, we'll go back to the barrel from *Chapter 1*, *Creating Low-Poly Models*, and create materials for it.

Creating materials

So far, you've been editing the default Blender material, but creating new ones is easy enough. We'll need at least one object that has some substance. You can either continue with the barrel you've designed or open the file in the `Start` folder in the `Chapter 2` folder of this book's GitHub repository. If you go with your own file, you'll most likely have the default material, labeled as *Material*, still in the **Material** panel. Using the minus (-) button, you can remove that and start fresh. The aforementioned file in this book's GitHub repository has already removed this default material for you.

It might be tempting to click the plus (+) button right above that minus (-) button you may have just clicked. Go ahead and do it. You'll end up with an empty line appearing in the material list. Those two buttons simply add and remove material slots to/from the objects, but not the materials themselves. Once you have a slot ready, you can designate a material for that slot. We'll investigate slots and different materials as we move forward, but let's create our first material by following these steps:

1. Select the *body* of the barrel.
2. Press the **New** button in the **Material** panel.
3. Change **Base Color** to a brown color.
4. Rename your new material `Wood` after double-clicking its title.

The following screenshot should help you compare your results with what's expected to happen. While selecting a color, you can hit the **Hex** button (shown in *Figure 2.1*) and type in `AD8654` so that your barrel is the same color as the one shown here:

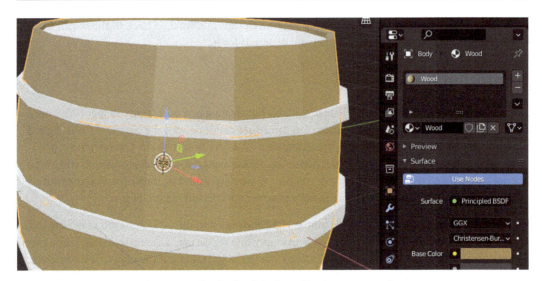

Figure 2.3 – The body of the barrel looks more wooden

Now, let's create another material – this time, a metallic one. But where did that **New** button go? In this situation, when the **New** button is missing, you can do the following:

1. Click the button with the plus (+) sign.
2. Press the **New** button.
3. Rename your new material Steel.
4. Pick an appropriate color for it, such as **555E64**.

So, this time, you have introduced a new slot and filled it with a new material. It would seem the **Body** object now has two materials and only one of them is in effect. Furthermore, we don't need that steel material for the body part of our model anyway. So, we should remove it. While the **Steel** material is selected, press the button with the minus (-) sign to remove it from the body.

Although you have created a new material and removed it, this doesn't mean it was all a waste. The material is still part of your Blender file but has been left unassigned. We'll make use of it soon. This means that you have used the **Material** panel like a workbench. Now, let's learn how to assign existing materials to objects.

Assigning materials

If you have your materials at the ready, then you can assign them to different objects easily. This saves you from creating the same materials repeatedly. We'll see how this is done in this section.

The **Ring** object has not been assigned any material, but you can assign the **Steel** material to it. Start by selecting the **Ring** object, then expand the drop-down menu next to the **New** button, as shown in the following screenshot. Select the **Steel** material from the list. This will assign the selected material to the object:

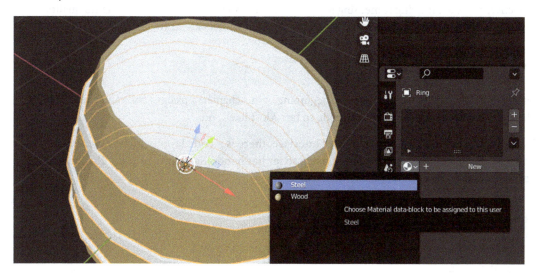

Figure 2.4 – Existing materials are listed in this drop-down menu

It's a good idea to name your materials according to their function, such as **Wood** and **Steel**, as it will be easier to find them later. You will also see that the color of the material is shown as a small icon next to the material's name; this helps to a certain extent, but it's limiting if you have many materials with similar colors.

The *lid* might use the same wood material but, maybe, we can change things up a bit. After selecting the **Lid** object, do the following:

1. Assign the **Wood** material.
2. Click the **New Material** button (it looks like two sheets of paper).
3. Rename **Wood.001** to `Dark Wood`.
4. Choose a darker color, such as **7E623D**.

You have just created a copy of the original **Wood** material for the **Lid** object. If you expand the materials list, you'll see all the available materials. Feel free to add, remove, duplicate, and/or assign as many materials as you wish for practice.

Additionally, you can assign materials not only to the whole object but also parts of that object. If you select some faces in **Edit** mode, then you can also apply a material to those selected faces only. In essence, the **Material** panel lists the materials associated with your model, regardless of whether it's applied to the whole model or parts of it.

Things must look a bit more colorful at this point. However, despite all your efforts, you can only go so far by just changing colors. The metal rings still don't look metallic enough. They should look more reflective, so we are missing something here. We need to discover new ways to give extra qualities to the base color. That's what shaders are for and that's what we'll tackle next.

Discovering shaders

Shaders were defined as two things at the beginning of this chapter: a piece of code and a user manual. Have you felt like you've been writing code so far? Most likely, no.

Nevertheless, behind that **Material** user interface, there is a code layer, which is the shader. For example, the default shader you've been using so far has hundreds of lines of code. The following is only a portion of the code that makes that shader:

```
metallic = saturate(metallic);
transmission = saturate(transmission);
float diffuse_weight = (1.0 - transmission) * (1.0 -
    metallic);
transmission *= (1.0 - metallic);
float specular_weight = (1.0 - transmission);
clearcoat = max(clearcoat, 0.0);
transmission_roughness = 1.0 - (1.0 - roughness) * (1.0 -
    transmission_roughness);
specular = max(0.0, specular);
```

Luckily for you, you don't have to write a single line of code. More importantly, Blender interprets the shader code so that it can offer UI elements such as color pickers to select colors, sliders to define a range of values, and drop-down menus that come with more advanced options so that you can utilize the shader easily.

The "user manual" aspect of shaders regards which properties of the material will be exposed to the user. For example, color is an obvious setting we should be able to change. The shader code will expose color and some of the other properties of a material to the outside world so that you can use the material easily. This is very similar to the way you use any device. You usually interact with a device via an interface by clicking buttons and turning some dials. A combination of these actions triggers certain events internally, which are not revealed to you, but you get to experience the result of these operations.

Going back to the original definition, you work with materials via a shader. These two go hand in hand. Moreover, just as Blender introduces a default material, it also comes with a default shader assigned to this material. It's called **Principled BSDF**. You can see this name next to the **Surface** section of the material's details. If you click **Principled BSDF** (**Principled** from now on, for simplicity's sake) in the interface, you'll see a list of other shaders. Selecting a different shader from that list will associate a different shader with your material. Some of the other shaders in that list are as follows:

- **Diffuse BSDF**: A basic shader that is responsible for displaying color on a surface. When objects are supposed to have a simple color – in other words, diffuse a certain color – this is the right shader to use.

- **Emission**: If you are designing an object and you want it to act like a light source, such as a fluorescent light, you can use this shader so that it looks like it's glowing.

- **Glass BSDF**: A shader with which you can simulate a glass surface. It comes with an **Index of Refraction** (**IOR**) setting so that you can decide how transmissive the glass is since there are different types of glass out there.

- **Glossy BSDF**: This is used to add reflection, which is great for simulating metals or mirrors.

- **Toon BSDF**: When you need the surfaces and edges to have a cartoony effect.

When you want your objects to show different qualities, then you will want some of the simple shaders to work together. For example, in a lot of science-fiction movies that show advanced machines and such, it is common to see glowing force fields that are also transparent. If you use **Glass BSDF** only, you'll see through but without a glow. If you use **Emission**, there will be no see-through visibility.

So, the **Principled** shader is the best combination of commonly used shaders. It acts like an uber shader that employs properties of different shaders under one roof. For that reason, at this point, it's best to stick with the default shader.

BSDF

You'll notice that some shaders come with this abbreviation. BSDF is a technical term and stands for **bidirectional scattering distribution function**. It is composed of BRDF and BTDF, which are responsible for reflecting and transmitting the light, respectively. Altogether, this system is responsible for how realistic the light will interact with your object. In layman's terms, it calculates how much of the light is soaked by the material, and how much of it will be reflected by considering intensity, angle, and so on.

Now, let's learn how to make modifications to the **Steel** material for our barrel. Unfortunately, there isn't just one setting you can turn on to give a surface a metallic look. It turns out that not all metals are created equal. Some metal surfaces look more reflective or shiny, while some look rougher, and so on. We'll use a mixture of the following properties with different values to get the result we want:

- **Metallic**
- **Specular**
- **Roughness**

The dictionary definitions of these words might be good enough. That being said, in the context of Blender, those three properties work in tandem to create different metal surfaces. Therefore, you need to balance the intensity for each, similar to working with a recipe sometimes. When you are changing these values, to see the effect, you need to be in **Rendered** mode. You can switch to it by pressing *Z* and then *8*.

> **Shader values**
>
> The numerical values you change for a shader don't have units but act more like a percentage or intensity. 0 means you want none of it. 1 means full scale. So, 0.5 means 50%.

Let's analyze the results shown in the following screenshot. A default Blender material comes with **0**, **0.5**, and **0.5** as values for **Metallic**, **Specular**, and **Roughness**, respectively. So, the sphere in the top-left corner has values very close to a default Blender material. This means that, by default, your models will have some shine and roughness:

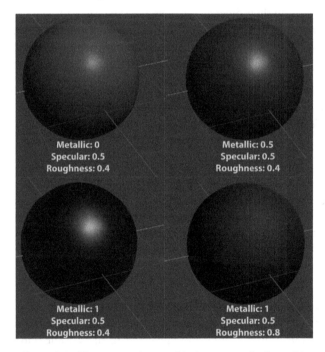

Figure 2.5 – The same spheres with the same color but with
different metallic, specular, and roughness values

In the top-right corner, you can see that only the metallic value has been increased. Even though the specular value is the same, we get to see more light being reflected. This makes sense because metal surfaces reflect more light. So, a surface that has more metallic qualities should reflect more light. This is exactly the case for the bottom-left model.

Finally, the sphere at the bottom right is what happens when we have a fully metal surface with an amplified roughness value. Notice how the shine is distributed more evenly on the sphere's surface because it's rougher. When light hits a rough surface, all the nooks and crannies of the surface will reflect much of the light in many different directions. When the surface is less rough or more polished, the light will directly bounce back into your eyes – in this case, Blender's camera. Hence, it will look shiny.

By the way, in all these cases, the base color is still the same, but the final look sure does feel different. The user manual page for the **Principled** shader contains a few charts that depict how some of this shader's settings interact with each other. It's similar to what's shown in the preceding screenshot but it comes with a lot more cases: `https://docs.blender.org/manual/en/latest/render/shader_nodes/shader/principled.html`.

The following screenshot shows the difference between two materials that both use the same base color:

Figure 2.6 – As expected, the metal rings are reflecting some of the light

The barrel on the left uses the default metallic, specular, and roughness values Blender provides. The barrel on the right has a material with **1.0**, **0.5**, and **0.2** set for its **Metallic**, **Specular**, and **Roughness** values, respectively. In summary, chances are you'll have to play with all three properties to get the metal look you want.

> **Non-metallic objects**
>
> Playing around with these three properties is also true for non-metallic cases, such as bricks, liquids, grass, and more. Let's compare a brick and a liquid, for example. Both can have the same base color – that is, blood red or some other tone. A brick is not a reflective object, so it should have very low – perhaps 0 – metallic and specular values. Most likely, its roughness value will be high. On the other hand, the liquid will need to be less rough and have a higher specular value.

Playing with the properties of a shader can be fun but it can also feel like you don't know what you are doing half the time. There is nothing wrong with experimenting to achieve the look you want. If you want to feel more confident about what you're doing, you can start observing objects around you. This may give you a better insight into choosing the properties that will give the result you've imagined. The property names are helpful in that sense, but they rarely work alone, so mix-and-match is necessary, even for professionals.

We'll investigate materials and shaders again when we cover Godot Engine, but here, you've seen how they work in Blender. Let's summarize what we have learned so far.

Summary

Throughout this chapter, you learned how materials can be used to give objects a different look. To create materials, you used the **Material** panel as a workbench to prepare many materials at once, and later assign these materials to different objects.

Shaders are almost inseparable from materials, and you got a glimpse of how many options they come with. You also saw that you can pick different shaders for your materials. However, most of the time, Blender's default shader, **Principled BSDF**, will be enough.

Using the default shader, you created a few materials that have different qualities, such as wood and steel. Furthermore, you discovered ways to create different-looking metal surfaces by utilizing metallic, specular, and roughness properties with varying intensities.

There is another topic that is usually covered alongside materials and shaders: textures. It was intentionally omitted, but it'll be covered in the next chapter with an explanation of why. For now, all that matters is that textures are graphic files that may enhance a material's visual impact. When you are ready, turn to the next chapter so that you can get to know them better.

Further reading

Blender's official documentation provides a detailed enough explanation of how different shaders and their properties work. The following URL lists many shaders you can investigate: `https://docs.blender.org/manual/en/latest/render/shader_nodes/shader/`.

Sometimes, seeing more examples may help you in your creation process. BlenderKit is a useful Blender add-on that you can use to access a whole bunch of materials and a lot more, such as models and scenes. Visit `https://www.blenderkit.com/` to read the installation instructions.

Since this book is about game development, we are only covering the basics of Blender in the context of helping us build a game with low-poly models. This means we are also limiting the level of detail that's employed while creating materials for the game. However, many professionals use Blender for different reasons, such as to create marketing material, product visualization, animation, and more. So, should you want to dig deeper into creating materials in a different workflow, here are some of the many great online courses out there:

- `https://cgcookie.com/course/fundamentals-of-blender-materials-and-shading`
- `https://www.udemy.com/course/become-a-material-guru-in-blender-cycles/`
- `https://studio.blender.org/training/procedural-shading/`

3

Adding and Creating Textures

In a typical 3D workflow, one of the most common properties you would add to a material is texture. A **texture** is an image file that is responsible for the textured look of a model so surfaces don't show just flat colors. Although objects you come across in real life have a perceived color, they also have a characteristic look that is defined by this property in 3D applications. For example, both a flower and a sandy surface may have a yellow color, but you know a flower's petal would look smoother, whereas grains of sand would look gritty.

Most day-to-day objects have wear and tear. Look around and you'll see that most surfaces will either have chipped paint, a slight deformation, or some scratches. Imagine the barrel you designed in the first two chapters has been in use for some time. It'd naturally have a few scratches on the metal rings. You can only go so far by applying colors to your materials and altering the roughness values. If you want to achieve a more realistic look, you've got to apply textures to your models.

Some 3D professionals only focus and gain expertise on certain domains. Texturing is one of these domains besides modeling, lighting, and animation. Typically, a texturing specialist will employ the help of classic image editing applications such as *Adobe Photoshop*, *GIMP*, and so on to create textures. Then, the artist will bring these textures into Blender so that they can be applied to surfaces. If you are not skilled in creating textures from scratch, you will learn in this chapter how you can still rely on existing textures out there created by other artists.

Preparing and using textures with the aforementioned workflow often sounds static because you need access to the source file of these textures. Luckily, there is a dynamic way to create your own textures within Blender, so you don't have to go back and forth between Blender and other software.

This is not a "one is better than the other" situation because each method has its own place and merits. You'll get to know new parts of Blender to facilitate both methods so you can make an informed decision about which texturing method to use. To that end, we are going to cover the following list of topics:

- Understanding UVs and texture coordinates
- Using the UV Editor

- Importing and applying a texture
- Creating textures procedurally
- Exporting your textures

By the end of this chapter, you'll have learned how to prepare your models for texturing, apply available textures, and create your own textures dynamically. The practice you'll gain in this chapter will give you insight into choosing the right method of texturing for your projects.

Technical requirements

This book's GitHub repo (`https://github.com/PacktPublishing/Game-Development-with-Blender-and-Godot`) will have a `Chapter 3` folder with `Start` and `Finish` folders in it for you to compare your work with as you go. These folders also contain other dependencies such as the texture files necessary to follow and complete the exercises.

Although you worked on a barrel in the previous chapters, we'll only use the standard Blender objects, such as a cube and a plane, to keep things simple so you can focus on the texturing workflow.

Understanding UVs and texture coordinates

While you are modeling, you are altering the coordinates of the vertices of a model. Thus, you are working with spatial coordinates. To apply a texture over your model, you need to work in a different kind of coordinate system that is called **texture coordinates** or **UVs**. Let's see how these two terms relate to each other.

The spatial coordinate system is often described with the **XYZ** acronym since we often use X, Y, and Z axes to define the position of 3D objects. Similarly, **UV** is another acronym but it is used in the texturing workflow; the letters U and V were picked to describe the texture coordinate system. So, UV doesn't really stand for ultraviolet.

The process that maps UV coordinates to XYZ coordinates is called **UV unwrapping**. Via this method, you tell Blender how a graphic file is mapped to XYZ coordinates. If unwrapping sounds counterintuitive, you could try to reverse the process in your mind. What kind of texture would you need so that if you wrapped it around your 3D model, it would fit perfectly? Let's analyze the following figure where a graphic file that is painted with a checkerboard texture is applied to a standard cube:

Figure 3.1 – A 2D checkerboard texture wrapping a 3D object

In *Figure 3.1*, you see a cube with a checkerboard texture on the left. In the middle part, you see the cube as if gift wrap is being peeled off. Finally, the cube is fully unwrapped on the right side; its texture is laid flat. The texture file is actually all of the checkerboard parts, and it exists as a 2D graphic file.

The reason we are using words such as unwrapping and 2D graphic files is because we are simulating a real-life 3D object on a flat screen. In reality, that cube would occupy a space, have a volume, and it would be full of the material it was made of. For example, a cube that might be a child's toy made of wood. Or, it might be a six-sided die, most likely made of acrylic. If you cut into it, you'd see the material.

To change the nature of the problem from a 3D volume problem to a 2D graphics problem, you need a new tool. You've been working with Blender's default interface, which is conveniently set up to edit XYZ coordinates. For editing UVs, you need the **UV Editor**, which you will discover in the following section.

Using the UV Editor

Blender comes with preset workspaces so you can focus on a particular workflow. So far, you've been in the **Layout** workspace. You can see it as the active tab just under the header of the application, next to the **Help** menu. You should create a new file and switch to the **UV Editing** workspace by clicking the appropriate tab. *Figure 3.2* is what you'll see when you are in the **UV Editing** workspace.

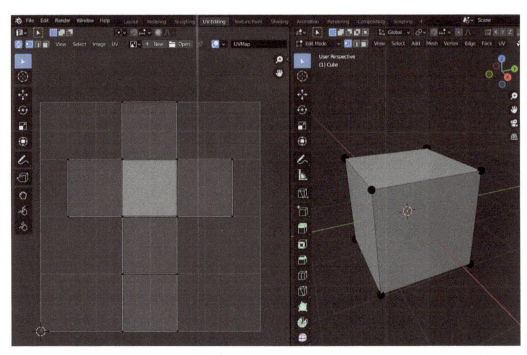

Figure 3.2 – UV Editing is one of many default workspaces in Blender

In the **UV Editing** workspace, the application will mainly be divided into two sections: the left side, which is called **UV Editor**, shows a bunch of squares laid out on a flat surface, and the right side shows the default cube. The black dots you see in **UV Editor** are actually the vertices of the cube in **3D Viewport**. You might notice that if you counted the dots in **UV Editor**, they don't add up to the number of vertices the cube has. There are more points in **UV Editor** because some of those points will eventually merge once those squares in **UV Editor** are folded around the edges and wrapped around your 3D object.

At this point, all of the vertices of the cube should be selected for you by Blender. However, if you happen to select a vertex of the cube, you'll see that the squares in **UV Editor** will disappear. That's because we haven't turned on the **sync** mode yet. At the top-left corner of **UV Editor**, you'll see a button with an icon that looks like two diagonal arrows going in opposite directions. If you have that button pressed, you'll notice that selecting the vertices in either view will synchronize.

When you add a new cube, Blender unwraps that cube by default. The general layout of the vertices in **UV Editor** resembles a T shape, like what you saw in *Figure 3.1*. Similar to **3D Viewport**, the vertices in **UV Editor** will form edges and faces, but it's all 2D in **UV Editor**. As mentioned earlier, we have converted the 3D-ness of the model to a 2D representation so we can work with graphics files.

UV Editor is where you can see how the points in the editor map or correlate to a texture file. To do that, we need to bring a texture file as follows:

1. Open the `Chapter 3` folder.

2. Open the `Start` folder.

3. Drag and drop `pips.png` into the **UV Editor** area.

If you open that PNG file in your computer's default image viewing application, you'll notice that it has transparent parts. Its dimensions of 1024x1024 are not fully painted. It just happens that the file's non-transparent areas come right under the faces in **UV Editor**, therefore the faces in **3D Viewport**.

> **Powers of two**
>
> Sooner or later, you'll notice that most texture files come in certain standard dimensions such as 512, 1024, 2048, and so on. Although these files don't have to be square, which means you could actually have 256x512 as dimensions, it'd still pay off to keep either dimension in powers of two. This is due to algorithms that are employed by GPUs so that they run more efficiently.

So far, we have taken advantage of Blender's default UV layout for a cube and have seen how UV faces can overlap with the texture file we have been previewing in **UV Editor**. However, if you enable **Material Preview** in **3D Viewport**, you won't see the die texture applied to the cube. That's because we haven't yet told Blender to use the die texture in the material assigned to the cube. Let's do that in the following section.

Importing and applying a texture

When you've dragged the texture file into **UV Editor**, you have effectively imported it, but, in reality, the material for the cube doesn't know how to use that texture yet. That being said, the material has all of the information it needs to map 3D vertices to 2D texture coordinates thanks to **UV Editor**. It just needs to be told which texture to apply to the cube.

To accomplish this, we'll switch to a new workspace so we can connect textures with materials. Also, we'll import another texture using a different method and assign it to the cube's material to showcase how you can use the same UV information with different texture files.

Just like when you switched to the **UV Editing** workspace, it's now time to switch to a different workspace for convenience. The sixth workspace, labeled as **Shading**, is the one you are looking for. We'll do our work in the lower half of the new workspace, which looks like a grid; it's called the **Shader Editor**. The upper part is still the same old **3D Viewport**, but **Material Preview** is automatically turned on so you can see your changes reflect immediately. So, the **Shading** workspace should look similar to what you see in *Figure 3.3*.

Figure 3.3 – Shading is one of many convenient workspaces set up for you

As you discovered in *Chapter 2, Building Materials and Shaders*, Blender files come with a default material. We'll modify that default material to understand the texture workflow. The **Shader Editor** area is already populated with two entities that make up the material as follows:

- **Principled BSDF** (**Principled** in short form)
- **Material Output**

These are called nodes. The node on the left, **Principled**, holds the properties you already saw in the previous chapter. A lot of these properties have little circles on the left side. These circles, which are called sockets, can connect to other nodes' sockets. We don't have enough nodes to create meaningful connections yet but we will soon.

Speaking of connectivity, **Principled** has an output that is connected to the **Material Output** node. If you hold your mouse down on the **Surface** input of **Material Output** and drag the connection away, you'll eventually break the connection between those two nodes. Then, the cube will look black since there is no surface information. Try to reconnect those nodes by dragging the **BSDF** output to the **Surface** input. The default gray color will be reestablished.

Nodes vs code

In the previous chapter, you were told that shaders are lines of code that instruct the GPU what to display. When you use nodes in **Shader Editor**, you are actually writing code, but you are coding visually. As the order of lines is important in conventional programming, the nodes and the connections coming in and out of the nodes are also important. However, visual programming is easier to conceptualize.

When we were modeling the barrel in *Chapter 1*, *Creating Low-Poly Models*, we needed to add 3D objects to the scene. We did that by pressing *Shift + A*. We'll do something similar. In this case, we'll add new nodes to **Shader Editor**. Blender is context-sensitive, which means the same shortcuts will yield similar results if your mouse is over different workspaces, areas, and interfaces. If you press *Shift + A* over **Shader Editor**, you'll see a list come up and show entities that are relevant to **Shader Editor**.

When this pop-up menu opens, it's positioned exactly so that the mouse cursor is right over the **Search** button. To add a texture node, perform the following steps:

1. Click **Search** in the **Add** menu.

2. Type Image with your keyboard.

3. Select **Image Texture** in the filtered results.

4. Click anywhere near the other nodes.

This will introduce an **Image Texture** node to **Shader Editor**, just as you see in the following figure:

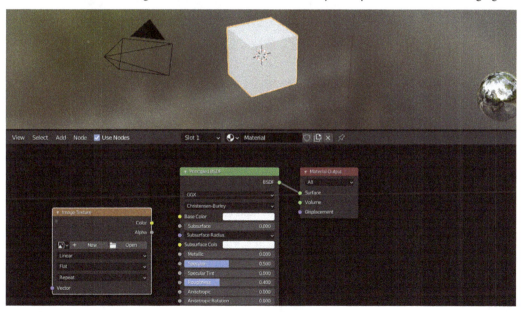

Figure 3.4 – An Image Texture node in Shader Editor

You have already imported the `pips.png` file when you were working with **UV Editor**, so there is no need to import that file again. We'll just recall it. As usual, the button to the left of the **New** button in the **Image Texture** node will bring up a list; select **pips.png** from that list. Then, attach the **Color** output of **Image Texture** to the **Base Color** input of **Principled**. This will apply the texture to the cube's faces. Voilà, the default cube now looks like a six-sided die as seen in *Figure 3.5*:

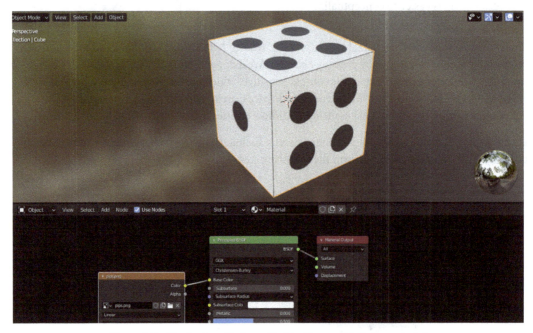

Figure 3.5 – The texture file is applied to the model via its material

A six-sided die has pips, usually marked with a variable number of circles on each side. What if you wanted to have a different looking six-sided die, with the numbers represented by Roman numerals? To import and apply a new texture, perform the following steps:

1. Create a new **Image Texture** node with the help of *Shift+A*.

2. Click the **Open** button.

3. Select `roman.png` in this chapter's `Start` folder.

4. Connect this **Image Texture** node's **Color** to the **Principled** node's **Base Color**.

Since the texture coordinates are already mapped in **UV Editor**, you can easily swap textures that have similar shapes with different designs.

When you work with more complex models, you've got more work to do in adjusting the UVs; as long as the UV coordinates are aligned with the right parts of the texture, you're good. However, imagine a different scenario. How would you go about modeling surfaces that look like they are showing a repeating pattern with slight deviations? In the following section, we'll look into a different texture workflow.

Creating textures procedurally

The word "**procedural**" has been used a lot in recent years, especially in the video game industry, to describe different things. Although one might say everything we have done so far is following a certain procedure, the word means something else in our context. When we imported the texture file in the preceding section, it was already designed for us. In other terms, it was a static file. The word "procedural," on the other hand, is a fancy word that means dynamic.

In a dynamic or procedural texturing workflow, the goal is to expose certain parameters of the texture so that the texture can be changed on the fly, instead of going back to a graphic editing application. Since it's all dynamic, you won't have to import graphic files, and you'll be able to change aspects of the final texture. For example, if the six-sided die was using a procedural texture, it'd be like changing the color and/or the size of the pips.

Procedural textures have another benefit besides their dynamism. Static texture files would need you to do the prior UV work so that the vertices would be aligned with the parts of the texture. In a procedural workflow, the pattern in the texture might be seamless, so you don't need to worry about the UVs. Seamless, in our context, means that the pattern repeats in a perfect way to wrap around the model.

We'll create a procedural lava texture as you see in *Figure 3.6* in Blender so you can change its parameters to have a different looking texture.

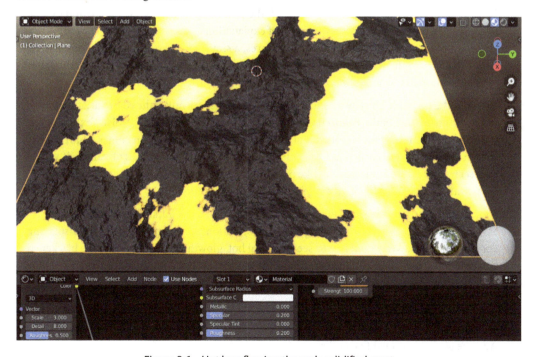

Figure 3.6 – Hot lava flowing through solidified crust

In a new Blender scene, after deleting the default cube, perform the following steps:

1. Add a **Plane**.
2. Switch to the **Shading** workspace.
3. Bring up the default **Material** or create a new one.
4. Rename the material if you desire.

    Nothing new or exciting so far, but we'll utilize the following five new nodes very soon:

    *   **Noise Texture**: Perlin noise is a mix of black and white values that are mixed together in a gradual way, so the result looks like a soup of grayscale values. Blender's noise texture is similar to Perlin, but the values are not grayscale; they come with random colors.

    *   **Bump**: It is used to simulate height fluctuations so surfaces could look bumpy.

    *   **Color Ramp**: Another name for this node would have been color mapper, but since it's using a gradient, the word "ramp" implies that the transition is smooth.

    *   **Emission**: Under normal light, hot objects have a glowing effect. This shader would help you simulate a hot piece of steel coming out of an oven or a bright lightbulb.

    *   **Mix Shader**: It's a shader that mixes two shaders to create a combined shader.

    Before we move on to how to mix and match the preceding list of nodes, which kind of look like a recipe's ingredients, here is a little bit of explanation as to why they were chosen. When you want to create your own procedural textures, a similar process might help you pick the nodes that are helpful instead of making wild guesses about which nodes to select. Also, after the explanation, try to imagine which one will connect to which. So, here we go.

    Noise Texture is quite literally a texture that comes with some noise; the color variation in this noise texture is used in the **Bump** node to simulate different heights. So, **Noise Texture** is like the data and the **Bump** node is its visual representation in a sense. Then comes **Color Ramp**, shown as **ColorRamp**, which assigns color information to different height values. If you've ever seen a miniature landscape, it's like painting hilltops white because of snow and the lower areas with different shades of green depending on the elevation.

    Hence, the first three nodes are taking care of most of the work for simulating elevation. Let's assume this lava texture is portraying a recent formation, so we are not after just displaying cooled-down lava. We would like to see steaming hot, glowing lava in between the blackened and dried-out lava. So, we'll need an **Emission** shader for that. Finally, since the elevation is its own thing and we are adding the emission part, we'll need **Mix Shader** to combine both.

    While working with nodes, you can drag and drop the nodes to arrange a cleaner layout for yourself to make sense of what's going on. Without further ado, let's continue.

5. Add the aforementioned five nodes.

6. Connect as follows:

    - **Noise Texture's Color** to **Bump's Height**

    - **Noise Texture's Fac** to **ColorRamp's Fac**

    - **Bump's Normal** to **Principled BSDF's Normal**

    - **ColorRamp's Color** to **Mix Shader's Fac**

    - **Principled BSDF's BSDF** to **Mix Shader's** first input **Shader**

    - **Emission Shader's Emission** to **Mix Shader's** second input **Shader**

    - **Mix Shader's Shader** output to **Material Output's Surface**

There is no left or right direction when it comes to connecting nodes. Some people consider a group of nodes as a unit and arrange them close to each other. So, sometimes, the last output node from that group connects almost vertically to another group of nodes. That being said, having a general flow of left to right would fit the preceding instructions. Whichever way you arrange your nodes, the layout might resemble what you see in *Figure 3.7*.

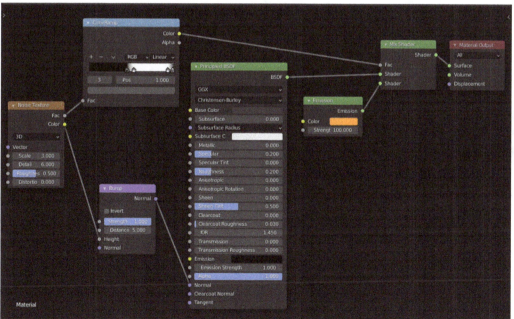

Figure 3.7 – Lava texture's node arrangement

Let's look at the values these nodes will have by following the original order of the node list as much as possible.

Noise Texture

For **Noise Texture**, the following values were used:

- **Type** defines the dimensions that are used in the creation of the noise, which involves complex operations. It's used in more advanced cases, so we'll leave the default **3D** value.

- The **Scale** property works more like a zoom factor. Too low, and you are closer to the noisy surface. Too high, and you are seeing a larger portion of the noisy landscape as if you are climbing up in an airplane. In this case, we set **Scale** to **3.0**.

- The **Detail** property is self-explanatory. Although having a lower value will certainly result in a muddy look, having a higher number beyond a certain value won't add much to the quality. It will simply increase the calculation time. A value of **8.0** is chosen in our case.

- **Roughness** is not the same concept you saw in *Chapter 2, Building Materials and Shaders*. That one affected the reflective properties of a surface. This one is about how rough the edges are, in a sense. In other words, how roughly the noise values are blending into each other, and a value of **0.5** is enough.

- The **Distortion** property creates swirls and wavy patterns. Perhaps a little might be necessary for a flowing lava look. You could experiment with it, but beyond a certain value when there is too much distortion, things will look too fragmented. So, **0.2** is good enough.

Bump

This node will use the data provided by **Noise Texture** so it can represent different color values as different height values. This is why the **Height** input was connected to the **Color** output since there can't be just one height value for the whole surface, so we had to feed it a set of colors.

Leaving the **Invert** setting unchecked, the following are the other values used:

- The **Strength** value determines the effect of the mapping between color values and the final bumps. It works like a percentage since the values can be anywhere between *0.0* and *1.0*. We'll leave it at **1.0**.

- The **Distance** property is a multiplier of some sort. It works in conjunction with the **Strength** property. Setting either one of them to *0* will result in a totally flat surface. Perhaps the best way to describe this property is that it keeps the details set in **Noise Texture**. Any value closer to *1.0* will show a washed-out surface, whereas higher values will fill in more details. Thus, a value of **3.0** will yield a detailed enough result.

Emission

This is a very simple node and it's responsible for making surfaces look glowing. We'll discover lights in *Chapter 4, Adjusting Cameras and Lights*, but if you want your objects to act like they are emitting or radiating light, then you can use this node. Examples might be a piece of hot iron or fluorescent lightbulbs; in our case, lava.

Since this is such a simple node, we have only the following two properties:

- The self-explanatory **Color** property is for picking which color the surface will emit. For hot lava, you can switch to the **Hex** values on the interface and choose **FF8400**.

- The **Strength** value, which is **100.0** in our case, defines the intensity of the emission. This is a unit measured in Watts so you can be scientific about it, but picking arbitrary values for visual fidelity works most of the time too.

ColorRamp

The **ColorRamp** node is used for mapping input values to colors with the help of a gradient that works like a threshold. The description is deceptively simple, but there is a lot going on under the hood. So, let's unpack it.

Most of the time, you'll be connecting both the input and output sockets of a node to other nodes. However, there are times when it is totally acceptable to use only one type of socket. For example, in the **Emission** shader, you didn't have to use the input sockets to define the **Color** and **Strength** values. Instead, you handpicked their values. So, the node acts like a source of information.

Then, there are some nodes where it makes much more sense to connect the input socket to another node's output socket. **ColorRamp** is one of those nodes, and it works like a modifier by factoring in incoming values. **Noise Texture**'s data will be a factor (Fac for short) in creating a lava surface, so we connect the two **Fac** sockets.

Once the data is factored in, we need a system to process it. This is done via the gradient in the **ColorRamp** node. The concept of a gradient might sound weird at first. If you were to connect the **Color** of **Noise Texture** directly to **Material Output**, you'd see that there are smaller and larger zones of colors. If you do that, remember to undo it so that the nodes are connected correctly once again. We need a way to turn these flat but colored zones to elevation.

The gradient is going to help us define which zones are higher or lower so we can assign the appropriate color to different elevations later. In essence, the gradient is a tool to define and blend in those zones with the help of color stops. By default, there are two color stops, but you can use the plus and minus buttons above the gradient to add and remove more color stops. These stops have a square shape with a little triangle right above them. It is possible to drag these stops, which will change the zone transitions we mentioned earlier.

When you have a lot of stops, it's sometimes difficult to click and drag them, so use **active color stop** to step between them. When you add a fresh **ColorRamp** node, the active stop is marked as **0** and it is to the left of the label that says **Pos**, which indicates the position of the active stop. Both the active stop and the position fields show necessary UI elements for you to change the values once you hover; also, you can click and enter a value. So, by using the active color stop and **Pos**, you can mark exactly where the color stops are going to be if you don't want to drag them around.

Last but not least, there is a color picker right above the **Fac** socket. You can use that to set the color for the active stop.

Since this is not a straightforward node, we could benefit from some visual aid. *Figure 3.8* is a zoomed-in look at the **ColorRamp** node.

Figure 3.8 – A close-up look at the ColorRamp node

The preceding figure should help you see what we have talked about so far. Also, just like you are able to zoom in and out with your mouse's scroll functionality in the 3D view, you can do so in **Shader Editor**. It will help you see some of the properties' names and values more clearly.

Now, it's time to use all of this information and mark our transitions; you'll be interacting with all of the elements just presented. To that end, perform the following steps:

1. Use the plus/minus buttons to have four color stops.

2. Set **active color stop** to 0, then do as follows:

    I.   Set **Pos** to 0.45.

    II. Set color in the **Hex** mode to 000000.

3. Set **active color stop** to 1, then do as follows:

    I.   Set **Pos** to 0.53.

    II.   Set color in the **Hex** mode to FFFFFF.

4. Set **active color stop** to 2, then do as follows:

    I.   Set **Pos** to 0.94.

    II.   Set color in the **Hex** mode to FFFFFF.

5. Set **active color stop** to 3, then do as follows:

    I.   Set **Pos** to 1.00.

    II.   Set color in the **Hex** mode to 636363.

Notice that we are only picking grayscale values. In a real landscape, higher areas will be cooler lava, and the lower areas will be hot pools of lava. So, to represent that idea, we are picking dark and white colors. Usually, the whiter something is, the hotter it is. The proximity of the stops to each other determines how smooth or sharp the transitions are.

Although we have been working with the **ColorRamp** node, the colors for our lava texture will be defined in the **Principled BSDF** and **Emission** shaders and will be combined in **Mix Shader**. For the time being, we have utilized the data from **Noise Texture** and transformed that data with the help of a gradient and its carefully chosen values. We'll revisit the factor concept again in the *Mix Shader* section, but before that, let's visit our trusty friend **Principled BSDF**.

Principled BSDF

We actually saw this node in *Chapter 2, Building Materials and Shaders*, but it was displayed as part of the **Material Properties** interface. When you create a new material, it uses this shader by default. It combines a great deal of other shaders in its body. For example, it has an emission socket, but since we can't do both the hot and cool part of the lava formation in one go, we are using a separate **Emission** shader.

We'll leave most options unchanged, but the following are the non-default values chosen for this exercise:

- **Base Color** is for picking the perceived color, such as green for grass and brown for dirt. You can set 4A4A4A as the value in the **Hex** section of the color interface.

- The **Specular** property defines the reflectivity of the surfaces. Since dried lava is not a reflective surface, we'll pick a small value such as 0.2.

- **Roughness** is for specifying how rough the surface will be. Although it sounds like a simple property, it's functioning in conjunction with **Base Color** and **Specular** values. So, picking an intuitive value is not always easy. You'd expect the dry lava to be rough, hence having a high roughness value, but we'll pick 0.2 in this exercise.

You can refer to *Figure 2.5* in *Chapter 2*, *Building Materials and Shaders*, and read the explanation in the *Discovering Shaders* section for a refresher in understanding how multiple properties work together and affect the final look.

Mix Shader

It blends one shader into another determined by the value in **Factor**. For the **Factor** socket's value, if you pick **0.0**, the first shader will be used entirely. If you choose **1.0**, it means that the second shader will be utilized.

The range of decimal values is between 0 and 1 but it's hard to know what to choose since we can't just arbitrarily determine how much of which shader to use. This is why we connected the **Color** output from **ColorRamp** as a factor so that the fluctuation in **Noise Texture** would trickle down and affect this node. The effect is cascading. In other words, every single pixel that's going to be painted either dark (for dried lava) or orange (for hot lava) should be decided based on where **ColorRamp** thinks it belongs in **Noise Texture**. Thus, the color stops act like thresholds and this is all factored in, in **Mix Shader**.

Once all of the nodes have been set and attached, feel free to play with the values in all of them, especially **ColorRamp**. You'll notice that the hot lava parts are sort of cooler at the shore, and denser and brighter in the middle. Try to approach the color stops close to each other and see how these hot zones in the lava pools change.

Creating this kind of texture using conventional image editing applications such as *Adobe Photoshop* might have been possible, but those applications are layer-based and it's not always easy to keep things non-destructive. The power you have with a node-based approach is quick iterations. One thing for sure is you don't have to reimport your texture to see the changes. It's all happening live in front of your eyes.

However, at the end of the day, since you are developing a game, you'll have to export your texture so the game engine of your choice can use it. In the following and final section, we'll see how we can export our lava texture to the file system.

Exporting your textures

In later chapters, when we get close to working with Godot Engine, we'll look into asset and project management in more detail. However, after all the hard work we have done with the lava material, it's now time to learn how to export the texture.

We'll do a few interesting but necessary things in this section to export our texture. First, we'll change Blender's rendering engine. Then, we'll add an **Image Texture** node in the middle of our material without connecting it to anything. Weird, right? Blender works mysteriously sometimes.

Changing the rendering engine

We have been using the default **Eevee** rendering engine so far. **Eevee** is a real-time rendering engine that gives you really fast results. Most game engines have their own internal real-time rendering engines that are responsible for calculating lights and shadows. So, **Eevee** is a good way to simulate in Blender what you'll most likely experience when you export your assets to a game engine. However, the speed and convenience come with a few penalties.

Blender has another engine that is called **Cycles**. **Cycles** is a very accurate but slow rendering engine compared to **Eevee**. **Cycles**' accuracy is due to the fact that it tackles advanced lighting calculations, which leads to quality results such as showing reflective and transparent surfaces much better, displaying more accurate shadows, and even creating volumetric effects such as haze and fog. The following is a link to an article that demonstrates both engines' capabilities and differences with use cases: `https://cgcookie.com/articles/blender-cycles-vs-eevee-15-limitations-of-real-time-rendering`.

In this book, we are not covering advanced enough topics that would require us to make a hard decision between **Eevee** and **Cycles**. So, **Eevee** has been fine for our purposes. However, when you work with procedural textures, there is no way, at least with the version of Blender we're using, for **Eevee** to export the lava texture. We'll have to switch to the **Cycles** engine. Luckily, it's done just with the click of a button.

In the **Properties** panel on the right, the second tab from the top, which looks like the preview display of a DSLR camera, is going to open **Render Properties**. The drop-down list at the top will show **Eevee**; let's change that to **Cycles**. Also, if you have a decent graphics card, you might want to change the third dropdown, **Device**, value to **GPU compute** so that your graphic card can do the heavy lifting instead of your good old CPU.

Looking down in that long list of properties, you'll see a panel with the header **Bake**. If you expand the header, you'll see a **Bake** button. We'll click that button soon, but we need to prepare what we'll bake first.

Baking a texture File

When we worked with the cube and die textures, we used an **Image Texture** node to bind an existing image from the file system. Our situation is different when the texture is procedural since this has been happening live in the memory. We need to figure out a way to bake this information into a file. Since there is no such file, we need to pretend that we have one, as follows:

1. Add an **Image Texture** node.
2. Click the **New** button.
3. Type `lava` in the name section.
4. Click the **OK** button.

We won't be connecting **Image Texture** to anything. If you remember the definition of what a material is from the early sentences in *Chapter 2, Building Materials and Shaders*, this new image we have just labeled as `lava` will be packaged with the material. Blender will make an educated guess and will bake the procedural texture parts into this image.

Now is the time to hit that **Bake** button in **Render Properties**. A progress bar at the bottom will indicate that Blender is doing its thing. Once the process is finished, the bottom-left corner of the **Shading** workspace will fill with the lava texture. That little section that displays the baked texture is called **Image Editor**.

If you look at the baked image, you'll notice that some details are lost. The pool of hot lava has warmer and cooler spots in **3D Viewport**, but the baked image has lost all of those details. This is because the **Emission** strength is so high that it saturates the baked image. It's like how digital cameras show poor-quality images when the scene is over-exposed with light. To alleviate this problem, you can bake again after setting the **Emission** node's **Strength** to `1.0`.

In the **Image Editor** interface, there is a button that has three stacked horizontal lines in it. If you click that button, you'll see a menu with two items: **View** and **Image**. If you expand the **Image** option, you can click **Save As** to save `lava.png` in your file system. This file can now be imported into a new Blender file and used in an **Image Texture** node. Then, you can apply this image in a material to a **Plane** object just like you applied the `pips.png` to a cube.

Mission accomplished. If you chose the same values as those written in this chapter, you should have the procedural lava texture you see in *Figure 3.6*. Additionally, you have created a static version of it. Let's summarize what else has been accomplished in this chapter.

Summary

This chapter started off with a brief discussion about what textures are and why they might be needed. To recap, if you are fine with models that have just the color info on their surface, you are done as soon as the modeling and material application process is finished. If you think you need to show distinctive qualities on your models' surfaces, you need to utilize textures.

To that end, you discovered how a new coordinate system—one that involves mapping spatial coordinates to texture coordinates via a method called UV unwrapping—might be necessary. Once the UV unwrapping is done, you can apply and swap different textures to your 3D models since the mapping from 2D to 3D is established.

Although creating textures with image editing applications is quite possible, you also know how to create textures procedurally in Blender. This is a powerful method, especially when it comes to surfaces that are hard to UV unwrap, such as landscapes.

Last but not least, you learned how to change the rendering engine to be able to export your procedural texture to your file system. Although this file is static and can no longer be updated automatically (unless you overwrite it with a new export, of course), you have the benefit of sharing the file easily.

You've been using Blender's interface and your mouse to move around the scene and rotate the view to have a better look at your models, materials, and so on. In the following chapter, you'll learn how to work with Camera and Light objects to create a composition where you can arrange objects in your scene under the best light conditions possible.

Further reading

To read more about what each shader node does, you can refer to the official documentation at the following link: `https://docs.blender.org/manual/en/2.93/render/shader_nodes/`.

For further practice, imagine where else the method for the lava texture could be used. Perhaps, with carefully planned values and more color variations, the hot lava might be rust, and the cool lava might be paint?

If you are curious and would like to investigate different software out there capable of producing procedural textures, you can give *Adobe Substance Designer* a try. It's a powerful program dedicated solely to creating textures. Not all of the nodes are labeled the same, but there are a lot of similar nodes to Blender's. In fact, if you practice your skills there and look at other people's creations, you might gain insight into creating such textures in Blender.

4

Adjusting Cameras and Lights

When you start a new scene, there are default **camera** and **light** objects in the **Outliner**. Although they are part of the scene, when you are modeling a new object, rotating around it, and looking at a material preview of it, you are still using Blender's internal camera and lighting system. This default behavior is good for working fast but doesn't produce artistic and accurate results.

In this chapter, you'll learn what a camera does and how to employ lights to get the look you want. The premise is simple: you can't see anything without a light, and you can't record or capture anything if you have no apparatus to do so.

Although it sounds like we are covering two distinct topics, we'll talk about both cameras and lights in this chapter. Between the two, we'll prioritize lights over cameras; you'll be provided with an explanation of why.

Thus, just like in real life, a camera and light conditions work together, and they go a long way to get the best shot you want. To that end, we will cover the following topics:

- Rendering a scene
- Understanding light types
- Introducing MatCap and Ambient Occlusion

After reading this chapter, you'll know how to pick the correct light type and capture a shot of your scene. You'll also know why you may want to postpone setting up cameras and lights. However, we'll offer you a way to attain some semblance of visual fidelity.

Technical requirements

We'll be entering new territory in this chapter, so it will be safer for you to rely on the files in this book's GitHub repository: `https://github.com/PacktPublishing/Game-Development-with-Blender-and-Godot`.

The appropriate filename will be mentioned when it's relevant. These files have already been set up for you so that you can focus on the material in this chapter.

Rendering a scene

In the computing world, the word **render** is similar to its other meanings in a dictionary. The rendering process in Blender will take a raw scene and produce a refined result. In more advanced cases, where your scene may have a **physics** or a **particle** object, this process will be responsible for calculating the state of these dynamic objects too. However, for brevity, we'll only look at what role the camera and light objects play in renders.

Let's create our first render by doing the following:

1. Start a new Blender scene.

2. Press *F12*.

Alternatively, you can use the **Render** menu near the application's title at the top. This should give you the following output:

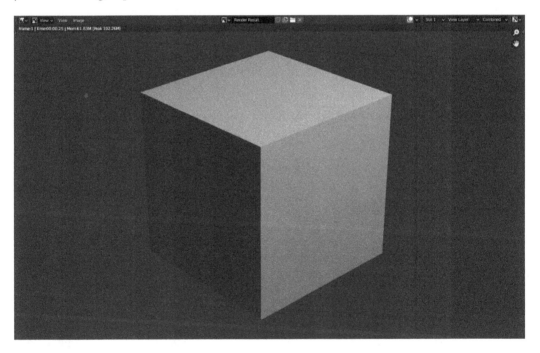

Figure 4.1 – Your first render of a default cube with Blender's default camera and light options

This is nothing exciting perhaps since this is pretty much the look you are used to seeing while working within Blender. The render is displayed in a separate window that covers the Blender window you were just working on. Therefore, you can close this window by pressing the operating system's close button or by pressing *Esc* to return to Blender.

If you take more renders and go back and forth, you'll notice that the grid underneath the cube and other objects, such as the camera and light, are no longer part of the render. This is expected. These objects, called **gizmos**, will facilitate things for you but won't be with you at the end of the journey. They work like scaffolding during the construction of a building. Although they are helpful while doing the work, they are taken away after the job is finished.

Let's repeat the previous exercise by changing one thing. What would happen if there were no cameras in the scene? Time to experiment:

1. Right-click **Camera** in the **Outliner**.
2. Delete this **Camera** object.
3. Press *F12*.

Did you expect to see a pitch-black render? Instead, you got an error stating that no camera was found in the scene. No camera means there isn't any instrument to render your scene, so Blender displays an error message.

Let's run a similar experiment by removing the **Light** object. After starting a new Blender scene, follow these steps:

1. Right-click **Light** in the **Outliner**.
2. Delete this **Light** object.
3. Press *F12*.

Let's speculate about what we expect to see. We have a camera to render the scene but no lights. Even though the cube object is part of the scene, we should not be able to see it. And yet, if you look at the following render, you will see a silhouette of the cube:

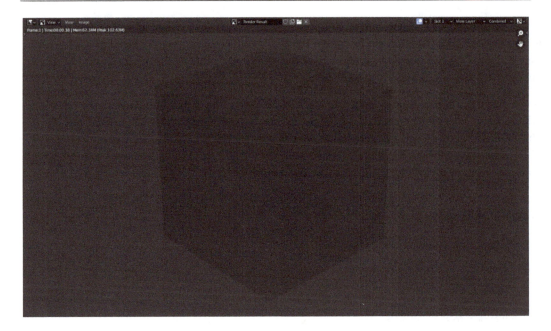

Figure 4.2 – An unexpected render when there are no lights in the scene

Most software applications come with default settings for the sake of helping out the user. In this case, Blender comes with a background color that contributes, unfortunately, to the result in the previous render. If you were to change the color of this setting to black, for example, then you'd have a completely black render. To achieve that, follow these steps:

1. Switch to the **Shading** workspace, as you did in *Chapter 3, Adding and Creating Textures*.

2. Switch from **Object** to **World** mode in the **Shader Editor** by using the dropdown near where the four views meet.

3. Change the **Background** node's **Color** property so that its **Hex** value is 000000.

The following screenshot shows the setup for changing the background color:

Figure 4.3 – We can also use the Shader Editor to change the scene's background color

If you take another render now, you'll notice that it's all black. There is neither direct nor indirect light or color contributing to the result. So, although things are looking rather dark, this is the result we expect to see. When would this be useful? If you would like zero surprises, which means you'd rather control every single light source and how much they contribute, then picking a black color for the background might be a good idea.

However, most Blender users are artists, not scientists. So, they often have multiple light sources and adjust these objects' settings to achieve visual fidelity, not scientific accuracy. Therefore, leaving the background color alone might be something you'll do as well.

Speaking of light sources and their settings, this is the right moment to segue into learning about the different types of light Blender employs. We'll light things up in the next section.

Understanding light types

So far, we have seen a render where the light object plays a role and another render when the light object was missing. We haven't discovered what this light object is. In this section, we'll get to know different types of lights. By the end of this section, you'll have a good level of knowledge of each type and why they matter.

We'll do this discovery in the context of the **Eevee** render engine because it simulates what game engines will do with your scene well. Since it's enabled by default, you don't need to make any changes at this point. Hence, you first need good knowledge of lighting your scene with the basic types of light. That's what we are going to do next.

Types of light

Let's look at the different types of light that are available:

- **Point**: This is the default light type you get when you start a new Blender scene. It's also called an omni light sometimes, short for omnidirectional, since it casts light in all directions. Lightbulbs are a decent real-life example of this light. Of course, in reality, lightbulbs don't cast light through their base but it's a good approximation.
- **Sun**: This type is used when you need a constant intensity of light. In other words, the light is so powerful that it doesn't lose any of its intensity along the way. Unlike the other light types, **Sun**, just like the Sun, also sends light rays in one direction only. Thus, the light rays are coming from an infinitely far away distance without losing their potency.
- **Spot**: When you need a flashlight-like light source, this is the light type you should use. It will emit a cone-shaped beam of light in the direction you point it. Most shopping centers and stores have lights of this type, usually hidden in the ceiling.

- **Area**: If you want to have a light source that has a large surface such as a window, TV screen, or office lights such as conventional fluorescent tubes, then **Area** lights are the way to go. You can also define the shape of the area. Since it is a considerably larger source of light in contrast to **Point** lights, the result, including the shadows, feels softer.

To get a much better feeling about what each light type does, you will open a file that's prepared for you so that you can quickly switch between different types of light. Follow these steps:

1. Open the Start folder inside the Chapter 4 folder. This can be found in this book's GitHub repository, which was mentioned in the *Technical requirements* section.

2. Open the Lights.blend file.

3. Hold *Z* and then press *8* to switch the visuals to **Rendered** mode.

The scene contains a cube and a large plane as a base to hold this cube. The four different basic light types are all in the same position, all with their default settings. Only the **Spot** light is enabled in the **Outliner** and you can see its effect in the following screenshot. By clicking the eye icon next to each light type in the **Outliner** back and forth accordingly, you can see what each light does. Notice the overall feeling each light creates by illuminating a certain spot or changing how the shadows appear:

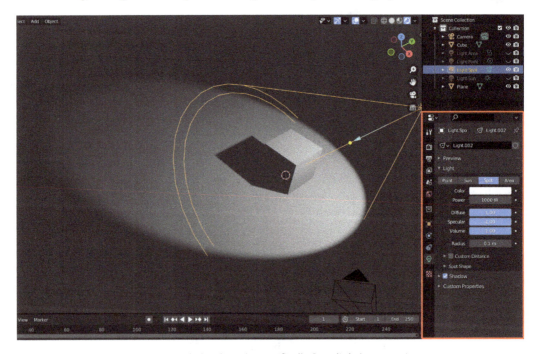

Figure 4.4 – A light object's, specifically Spot light's properties

Now that we have seen what each light does, let's learn about some of their properties.

Basic properties of light

The sample file was set up so that when you open it, the **Properties** panel should already be switched to the appropriate **Light** tab; this will display the five common properties that all the basic lights share:

- **Color**: This is the tint of the emitted light. If you are designing a fireplace, you may want to pick an orange or red tint, for example.

- **Power/Strength**: This defines how powerful your light source is in **Watts**. Thus, the higher the value is, the more powerful the light will be. In the **Sun** light's case, the **Power** property is labeled as **Strength**, but the concept is still the same. If you are designing a scene where accuracy is paramount, and you would like your lights to be as realistic as possible, then you are in luck. The *Power of Lights* section at the following URL lists values for some known light sources: `https://docs.blender.org/manual/en/2.93/render/lights/light_object.html`.

- **Diffuse**: In *Chapter 2*, *Building Materials and Shaders*, you worked with materials and set the color for the materials you applied to the barrel parts. The **Diffuse** property of lights works like a multiplier. So, keeping it as `1.0`, which is the default value, won't change the perceived color of a material. Decreasing it will diminish the color's effect on a material. In essence, this value determines the impact a light source has on a material's color.

- **Specular**: This is similar to the **Diffuse** property, except it affects the **Specular** quality.

- **Volume**: This is a bit of an advanced topic that involves more sophisticated settings when you set up materials. We won't cover advanced material settings in this book. However, like the **Diffuse** and **Specular** properties of lights, which work as multipliers, this property determines the light's contribution over a volume.

Out of these five properties, you'll most likely never touch **Diffuse**, **Specular**, and **Volume**. This is because, most of the time, it makes sense to change diffuse and specular values in a material. Also, volumetric light is an advanced case that can be handled via other means, similar to adjusting it via a material's properties.

More esoteric lights

If you are the curious type and read up on lighting, generally within the context of 3D applications, you will hear of terms such as **ambient light**, **global illumination**, and others. Even though those terms are relevant and important when producing a render, we won't cover them in this book for two reasons. First, basic light types are often enough because this will give you a more direct result and feeling for your scene. Second, the advanced lighting systems rely on and affect basic lights by making tweaks. So, understanding the basic types would be a better investment as a beginner.

Specific properties of each light type

Although you now have basic knowledge of what each light does, we haven't investigated what kind of setting contributes to the uniqueness of these lights. Now, let's look at each light's settings, which give the light its characteristic look and feel.

Point

Radius is a setting that's also used for **Spot** lights, but we'll cover it under this section since there is nothing else going on with **Point** lights. We've already considered a lightbulb as an analogy to **Point** lights. In reality, lightbulbs come in different sizes. So, you can imagine the radius value, measured in meters, as a mechanism to determine how big the lightbulb is.

The effect this value has is in the way the shadows are calculated. The default value, 0.1, will produce a rather sharp shadow. Try to increase this value to 1.0. You'll notice that there will be multiple shadows overlapping each other, following a direction away from the light source.

If you increase the radius to 10.0, something interesting will happen. The bulb is large enough that it will encompass the cube. It's so large that it intersects with the plane too. The shadows for the cube are no longer following a direction strictly away from the light source. The light source is so large it's as if there are multiple tiny point lights scattered inside a sphere with that radius value.

Sun

In some 3D modeling software and game engines, the **Sun** light is often labeled as **directional light**. There is a good reason for that. In real life, the Sun is so far away but so powerful that it's as if all light rays are parallel to each other. So, the **Angle** property defines the direction of the rays.

What about the position of a **Sun** light? You could try to move its location, but the overall effect on the scene won't change because the light rays are assumed to have constant energy, regardless of where they are coming from. So, the angle is the only meaningful factor for this light type.

Spot

A **Spot** light has the same **Radius** property as a **Point** light does. So, initially, they start as the same thing, then a **Spot** light sheds its light while following a conic shape.

There is a collapsed section labeled as **Spot Shape** in the **Properties** panel for this light type. This section houses two properties:

- **Size**: Measured in degrees, this value is the angle of the cone's origin. The higher the value is, the wider or larger the area will be once the light hits a surface. Similarly, lower values will focus the light on a smaller area.

- **Blend**: Once you have defined the area of light via the **Size** property, you'll have shaded areas outside the illuminated zone. The **Blend** value, which will be between 0.0 and 1.0, works like a percentage to adjust how smoothly these two contrasting zones blend into each other. Lower values will have a sharper transition. So, having it as 0 means a very sharp separation.

Area

For this light type to be more effective, deciding on its **Shape** setting is important. Four shapes exist:

- **Rectangle**
- **Square**
- **Disc**
- **Ellipse**

For all of these, you can customize the size of the shape. For example, the **Rectangle** shape will accept two values, but the **Square** shape will only need one dimension. You won't see much difference in the test scene if you play with different values. However, rest assured that they make a real difference in a much more complex scene where you distribute **Area** lights with different shapes.

Wrapping up

Adjusting light settings is only the beginning. Most 3D professionals dedicate themselves to certain disciplines. Lighting is one of these disciplines where you work on topics such as global illumination, bloom, volumetric effects, and many other advanced topics we won't be covering in this book. With that being said, using cameras and lights in Blender may still be useful to get a basic feeling about the artistic direction you are taking. For example, if you are designing a car, the headlights will most likely house a **Spot** light. If the model were a torch, a **Point** light might be appropriate.

Now, you may be thinking that we didn't cover a lot about lighting, but we also covered even less about the camera. This is because this book is about game development. In *Part 3, Clara's Fortune – An Adventure Game*, we mentioned that most of our work will be done in Godot, so you'll see that there will be many things we'll set up and fine-tune in Godot. Some of that effort will be for the camera and different light objects. Since we've been building individual models or constructing materials for our models, which will all be imported into the game engine in the end, there is no need to do a meticulous amount of work within Blender regarding cameras and lights. In other words, it's practical to set up cameras and lights in Godot because the settings in Blender won't transfer.

Now that you know why you should generally ignore Blender's cameras and lights, let's look at two helpful methods that will make your time more pleasant while still working in Blender.

Introducing MatCap and Ambient Occlusion

Since making more investment in a high-fidelity lighting setup in Blender no longer makes sense, we should perhaps investigate different ways to make our scenes look better. What we'll do next still means what you see won't be exported. However, it means you can look at models that no longer have the default and boring gray look. Why not? Working with things that look nice sometimes feels nicer and increases productivity. We'll look at two techniques that will help you distinguish your models' details.

MatCap

MatCap stands for **material capture**. We won't get into the technicalities of how a **MatCap** is constructed but, suffice it to say, it's a type of shader Blender uses internally to give a different look to the models. Normally, you'd need to switch to **Material Preview** mode to see how your materials would look on your models.

However, during the modeling process, you usually work in **Solid** mode because it's more performant for Blender to show you the changes you are making to your models. Thus, while still working in **Solid** mode, if you want to have a better visual as if you are in **Material Preview**, you can instruct **Viewport Shading** to use **MatCap**. So, it's the best of both worlds.

To make sure you are using **Solid** mode, do the following:

1. Press Z.
2. Then press 6.

This will switch **Viewport Shading** to **Solid** mode. It's also represented as a disc in the second icon at the top-right corner of the **3D Viewport**. We'll make some changes to **Viewport Shading** so that your models can have more pronounced details. If you click the down-looking arrow on the right-hand side of those icons, you'll expand a panel. This panel is shown in the following screenshot:

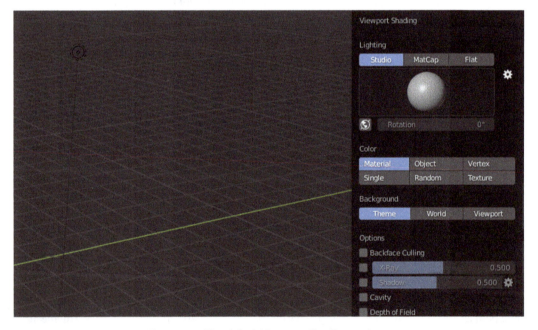

Figure 4.5 – The default Viewport Shading options

The settings in that panel let you change the way the models are displayed while you are editing them. You can already see a preview of the current settings as a sphere in the upper section. Let's click the second button, **MatCap**, under the **Lighting** title. This should already change the look of the preview in that panel, as well as the model's look in the scene.

We won't be discovering the **Color** part but try out the **Random** option for the barrel from *Chapter 1, Creating Low-Poly Models*. You'll see that different parts of the barrel take random colors. This helps to distinguish separate parts in your scene. Similarly, we will leave the **Background** setting set to **Theme**.

Let's investigate the **Options** section and focus on the parts that will give us a decent result:

1. Enable the **Shadow** option.
2. Set its value to 0.5.

You won't normally see the effects of the light sources in **Solid** mode, but the last operation will create a shadow effect. It's a cheap effect that efficiently creates depth.

Sometimes, your models will have parts that are away from the center of mass. These outer parts may also create areas that would look deeper from your point of view. Hence, you'll have cavities. To mark these areas more clearly, do the following:

1. Enable **Cavity**.
2. Set its **Type** value to **Both**.
3. Set **World Space** like so:
    I. **Ridge** to 0.5
    II. **Valley** to 1.0

4. Set **Screen Space** like so:
    I. **Ridge** to 0.75
    II. **Valley** to 1.0

This should create a big change in the way your models look. The **Cavity** option, with **Type** set to **Both**, will seek parts of your models that are at different elevation levels and accentuate them. In a way, if your model was laid out like a landscape, the ridges and valleys would be emphasized so that they would be more noticeable. The values we picked are a bit arbitrary, so feel free to alter them according to your taste or the complexity of the models.

Last but not least, in the settings for **MatCap**, if you wish, you can pick a different material. After all, we are still looking at a gray cube, even though we have improved its perception. For example, you can do the following:

1. Click the sphere preview under the **MatCap** button in **Viewport Shading**.
2. Select the third sphere in the second row.

If your version of Blender has the selection interface organized differently, we are looking for a sphere that looks like brown clay. This will change the look of your cube to, well, muddy clay. The following screenshot shows what we have done so far:

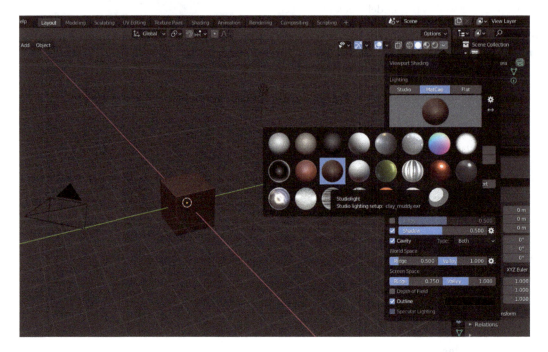

Figure 4.6 – Viewport Shading offers many ways to create a different look for your models

If the muddy color is too dark, then the second sphere in the first row is a nice alternative. However, keep in mind that this is only for you to feel at ease while working with your models in **Solid** mode. None of these changes will have any impact on the result when you render or export your models to other software. These are, in a sense, throw-away materials that will make your experience in Blender more pleasant.

So far, we have treated the **Solid** view as if it was **Material Preview**. This is useful when you want a bit more visual clarity without previewing the model's assigned materials since that makes extra calculations by taking into account the lights too. Next, we'll look into a way of doing something similar in **Rendered** mode.

Ambient Occlusion

In this section, we'll discover another handy visual tool that can help you have a bit more visual fidelity. This tool is called **Ambient Occlusion (AO)**, and it's also a method that's used in most games to create a more realistic look. Let's explore how and why this works.

Let's get the definition out of the way first. We have two names: ambient and occlusion. In the context of Blender, as you may have guessed, ambient is a term that's used to describe the overall light conditions. We switched the background color to black to modify the ambient light near the end of this chapter's *Rendering a scene* section. So, we are already familiar with this concept.

Occlusion means to obstruct or block something. In our context, it means to obstruct light. So, we want some light to be obstructed or occluded. But where exactly would we want this?

Take a look around wherever you are. You'll notice that some areas, by having a flatter surface, will be exposed to the natural or artificial lights coming off the ceiling or windows. Light – more specifically, the photons that make up the light – will be bouncing off these surfaces. Wherever these flat surfaces meet and make some sharper and some more moderate angles, they will be forcing the photons to scatter in a zigzag manner. As a result, it'll be harder for light to reach certain spots, so the geometry of your models is going to occlude some of the light.

To see the effect of AO, open any of the following files from this chapter's `Start` folder:

- `Lights.Area.AO.blend`
- `Lights.Point.AO.blend`
- `Lights.Spot.AO.blend`
- `Lights.Sun.AO.blend`

Also, remember to switch to **Rendered** mode by pressing *Z* followed by *8*. Otherwise, the effect won't be visible. Do you notice the darker part where the cube meets the plane? That's AO, as shown here:

Figure 4.7 – Ambient Occlusion visible where the cube touches the plane

The example files have been prepared so that the **Ambient Occlusion** option should already be visible on the right-hand side in the **Properties** panel. By switching it on and off, you can observe the behavior. AO affects the edges as if there is an extra volume of shadow, where shadows naturally would occur. This makes it look more realistic. We'll look at how to take advantage of AO as a separate effort inside Godot Engine later in this book.

Additionally, in the AO settings, if you pick a higher **Distance** value, it will sample a larger area from the object's contact zone. This may help you have smoother or sharper AO.

We've covered a great variety of topics in this chapter. Now, it's time to summarize what we've learned.

Summary

We started this chapter by rendering a scene with and without a camera and lights. During this effort, we utilized **Shader Editor**, which was introduced in the previous chapter to change the background color, also known as ambient color.

Then, we looked at different light types and how each type can be used to simulate real-life cases. We did this using the Eevee rendering engine. Should you switch to the Cycles render engine, the lights will have additional and more advanced properties, but the concepts you learned about in this chapter will hold.

We also discussed the fact that your rendering concerns will be left for later when we tackle things in Godot. However, it'd be a much more pleasant experience if we could work with better-looking things. To that end, you were introduced to two different methods.

The first method is **MatCap**, which you can use to change the way models look, despite not turning on material previewing. The second method, **Ambient Occlusion**, involves getting a feeling of where objects meet and how they behave under existing light conditions. You can use both methods at the same time if you wish.

In the next chapter, we'll move things a bit. You'll be studying and preparing a model for animation. For this effort, you'll utilize a process called **rigging** and simulate a skeleton-like structure inside your model so that you can animate it.

Further reading

Although this chapter covered cameras and lights, such topics are usually covered under the *Rendering* title in many publications. That's because there are different rendering engines, and each one treats lights and cameras differently. Also, **post-processing** and **color management** might be your concern if you want to take on more advanced renders. So, cameras and lights are only a small portion of the rendering process. To learn more, Blender's official documentation page might be a good start: `https://docs.blender.org/manual/en/2.93/render/index.html`.

Also, here are a few online resources that might help you dive deeper:

- `https://cgcookie.com/courses/fundamentals-of-digital-lighting-in-blender`.

- `https://cgcookie.com/courses/production-design-with-blender-2-8-and-eevee`.

- `https://cgcookie.com/courses/fundamentals-of-rendering`.

5
Setting Up Animation and Rigging

In *Chapter 4*, *Adjusting Cameras and Lights*, you saw why you should ignore certain concepts in Blender, specifically cameras and lights, because they don't transfer easily to Godot. This chapter is sort of an opposite case. You might be wondering whether a game engine can't move objects around for us, right? After all, we use a game engine to facilitate things such as displaying models, creating environments with visually rich effects, and so on. It's normal to expect a game engine to take care of animating our models as well.

Although animating simple objects is perfectly possible in Godot, doing it for complex models such as a human character (or any bipeds, such as a robot) or a lion (or any quadrupeds, such as a cow) will take a lot of effort. Therefore, it makes much more sense to do most animations in Blender because it offers a much more streamlined workflow. We'll explain in detail why that is so you can apply a similar reasoning process in your own projects.

Sometimes, you will have a model that looks nice and complete, but it won't be suitable or ready to be animated. In *Chapter 1*, *Creating Low-Poly Models*, we discussed vertices, faces, and edges. We will revisit some of those concepts in the context of getting our models ready for animation.

Then, when we believe the model is ready, we'll look at Blender's animation capabilities. We'll do this by discovering two new things. First, we'll utilize a new method called **rigging** and construct a rig that's ubiquitous in animating models. Second, we'll switch to a new workspace dedicated to animations. During this effort, you'll get to know a whole different side of Blender.

After you see how rigging is done and how models can be animated, we'll look into ways to prepare and store more animations in Blender so that they can easily be used later in Godot. So, once you know beforehand what will be required down the line, this knowledge might help you in setting things up accordingly in Blender before it's too cumbersome to change later.

Despite the following section titles looking deceptively short, we have a lot to cover in this chapter:

- Where to build animations
- Understanding the readiness of models
- Creating animations
- Getting animations ready for Godot

In the end, you'll know whether Blender or Godot is the right environment to tackle animations and how to get models ready for animations so that you can rig them.

Technical requirements

There will be a lot of moving parts, figuratively and literally, in this chapter. Animation and rigging are challenging topics for most people who start practicing 3D. Although we'll take things step by step, to give you extra help along the way, you might want to use some of the files that are in the interim stages instead of doing it all at once.

As usual, the book's repository will have the necessary files for this chapter at the following link: `https://github.com/PacktPublishing/Game-Development-with-Blender-and-Godot`.

Where to build animations

Both Blender and Godot Engine have animating capabilities. Therefore, you might be wondering which software is better for creating animations. To answer this crucial question, we should be discussing what we are animating. When it comes to animations, especially in game development, we will be tackling the following two main concepts:

- **Whole-body objects**: Objects such as a bouncing ball, a boat, or a projectile thrown from a source are all examples of objects that act like a solid system with no individually moving parts. The system can move as a whole without depending on its individual parts.

- **Connected systems**: Some systems depend on individual parts to be in motion. These systems have parts that are connected to each other and the individual parts work together to move the system they are part of. For example, cats use their feet, birds use their wings, and a human body moves in a certain direction using two appendages that are either in contact with a surface or interact with the medium they are in.

Sometimes, some tools and gadgets in real life can do a similar job, and it's possible to use one over another for a quick solution. However, every so often, we would like to pick the best tool for the job. We'll discuss both Blender and Godot in the context of the concepts we have just pointed out to see which option might be a better choice.

Animating in Godot Engine

Godot has a component, **AnimationPlayer**, that helps you build animations. We'll look at it more closely in later chapters when we import our models to create a point-and-click adventure game. Similar to other applications' animation components, it depends on setting **keyframes** to mark the changing points of an animated object. For example, to create a bouncing ball animation, you'd mark the ball sitting still on a plane in the earlier frames of the animation and mark a higher position in the world in the later frames.

This is quite easy to do with Godot. You just have to mark the important events as keyframes, and this operation is called **keying** or **inserting a key**. Thus, the engine figures out how the object should move in between the two keyframes. However, when the system is much more complicated than a simple ball, and it has moving parts, you'd be expected to select these separate parts to **key** them. This is not easy to do in Godot since the workflow is not constructed in a way to facilitate such complex operations in an easy manner. Consequently, it's best to use Godot Engine when the system is relatively simple.

Animating in Blender

As was just mentioned, when you are animating an object with parts that are responsible for creating the overall motion, such as animating a human body by moving individual parts such as feet and hands, then doing this kind of work in Blender will be the right choice thanks to a method called **rigging**. Later, in the *Creating animations* section, we'll explain what rigging is and discover how to construct a rig for our models.

For now, it should be enough to know that individually moving parts for an animated body will require rigging to expedite the animation process. This is where Blender shines because it offers tools and custom interfaces to help you along the way.

Besides the ease of creating an animation, let's point out another reason why Blender is a better choice for animating complex systems. If you construct your animations in Godot, you can only use them in Godot. On the contrary, a Blender animation will act as a source of truth so you can share it with other applications.

Wrapping up

We'll say one more thing about why Blender might be a better choice regardless of the complexity of creating animations. If you ever want to create a trailer for your game and you've gone through the trouble of creating accurate enough camera and light conditions similar to the ones you are going to employ in your game, then you can take a render of your scene, composed of many frames, which will utilize Blender's animation system.

So, for simple objects that can be moved, utilize Godot Engine's animation system. For systems that have individually moving parts, it's better to do it in Blender. After all, Blender has dedicated tools to facilitate the creation of advanced animations. Now, let's discuss when your models are ready to be animated.

Understanding the readiness of models

In *Chapter 1*, *Creating Low-Poly Models*, we started with primitive objects and altered their vertices, faces, and edges. During that process, we were concerned with how the model would look. As corny as it may sound, looks might be misleading sometimes. To be animated correctly, a model has to respect certain conventions other than how it looks. In other words, you've got to be sure whether your model is ready.

Topology and rigging

The readiness level of a model could be defined by the term **topology**, which sounds a bit technical. In layman's terms, it's the distribution and arrangement of the vertices, edges, and faces of a model that altogether mark how optimized the model is for animation.

Not all topologies are created equal. There are bad and good topologies. Let's look at *Figure 5.1* to get a better idea about what we mean by topology or distribution, particularly as being bad or good.

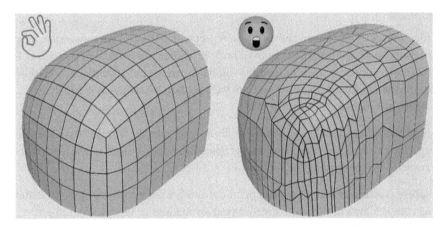

Figure 5.1 – The same model with two different distributions of vertices

The preceding figure shows a model with the same shape, but the topologies are different. Simply put, the left case is ready for animation, and the right one could use some work to straighten up those vertices to form a good flow. Then, you'd have to fix some of the irregularities by evenly distributing many of the faces that congregate. So, not only is the right case an eyesore, but it's also detrimental during the animation process.

Let's briefly touch on the role of rigging to understand the importance of good topology. If you were to model a human hand, you'd be designing fingers, knuckles, and the wrist. The model, or more correctly, its volume, would be hollow. In other words, you'd only be creating the vertices that would give the shape of a hand. However, in our minds, we know that this hand should have bones inside. When you wiggle your fingers around or bend your fingers at the knuckles and joints, different parts of the skeleton start moving so that the outer structure that's connected to the bone system can move accordingly.

To simulate this, you take advantage of a practice called rigging, which involves introducing a skeleton system and a series of constraints that manage how the skeleton system behaves. We'll work on a rigging example later in the chapter. For now, we are still concerned about our models being ready for the rigging to take place. To emphasize the relationship between topology and rigging better, let's turn our attention to *Figure 5.2*.

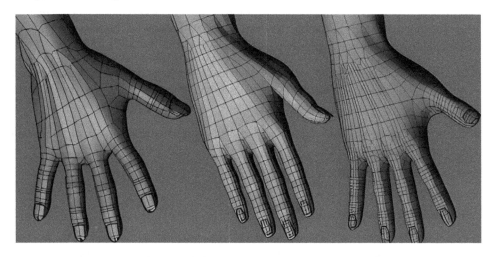

Figure 5.2 – Different topologies for a hand model

Observe how the faces are aligned more naturally in the middle case, which certainly looks like an improvement over the left one. Then, still for the middle case, look where the big thumb meets the main part of the hand; that area could use a bit more detail so that when the thumb stretches out like in the right case, there would be enough geometry to accommodate the skeleton's behavior. Compare the first and the third hands to see which one looks more natural to your eye when it comes to flesh and skin in between fingers.

When a model is bending or stretching at certain points, it will be creating some creased and protruded areas, similar to where the fingers meet the hand in the preceding figure. If vertices, hence faces, don't have a smooth flow, the model will look ripped or crushed in these weak spots. Having the correct topology is a topic that's hard to master and it throws off a lot of beginners when they want to get into animation and rigging. You can find a few links that can help you understand the difference between a good and bad topology in the *Further reading* section.

To satisfy a good topology, since it's necessary to line up edges and faces correctly where the action will occur, we need a mechanism to move problematic edges and faces around so that they will be in the right place. For this, we are going to discover a new method, or rather, a shortcut.

Grabbing

In *Chapter 1*, *Creating Low-Poly Models*, you got to know two methods that are very commonly used among Blender fans. They were **Rotate** (*R* as a shortcut) and **Scale** (*S* as a shortcut). There is a third common method that we intentionally omitted during that exercise. We depended on modifiers that helped us move vertices around, so we got away without it; however, it's now time to employ it.

If you are able to rotate and scale things, then why can't you move things around? In fact, you can, and this new method will help you move vertices, edges, and faces anywhere you want. There is only one caveat. Although most people refer to this operation as **Move**, its shortcut is a bit bizarre; it's *G*. So, an easier way to think of this shortcut in the context of moving might perhaps be grabbing. You grab a vertex and leave it somewhere, in a sense.

In most Blender tutorials, you may find people use grab and move interchangeably. They're one and the same. So, throughout this book, when you see the word move, we mean the grab operation and the *G* shortcut.

Let's practice this new piece of knowledge with a series of simple steps. After you start a new file, perform the following steps:

1. Press *Tab* to enter **Edit Mode**.
2. Select only one vertex of the default cube.
3. Press *G* and move your mouse around.

The vertex you selected is now being pulled around while you are moving your mouse. To terminate the grabbing, you can click anywhere and this should rest the selected vertex at its last position. *Figure 5.3* is an example of what we want to achieve.

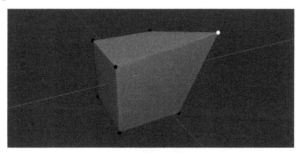

Figure 5.3 – A vertex grabbed out of its original place and moved somewhere else

You might have surmised that the vertex was moving freely in all three axes, and that would be correct. If you want to limit the movement to a certain axis, and if you wish to move the vertex a precise amount, you can do that too. While still in **Edit Mode**, perform the following steps:

1. Select another vertex.
2. Press *G*, then *X*.
3. Type 0.5.

You can pick any one of the other two axes if you want. Regardless, the value you type for any given axis defines the movement amount. So, a negative value will still move the selected part in the axis you choose, just in the opposite direction.

Additionally, sometimes you might want to move the selection in any but a certain direction. When you initiate a grab shortcut, if you press *Shift* before you pick the axis, it'll move the selection to the other two remaining axes. So, *Shift+X* would move things anywhere but on the X axis.

Practice the grabbing operation a bit more by selecting edges or faces if you would like. Soon, we'll explore the building blocks of animation. During that effort, you'll most likely utilize the grab operation. So, when you are ready, let's see how we can animate things.

Creating animations

As we mentioned in the *Where to build animations* section, the type of animation we'll do in Blender involves having individual parts of a system that move independently from each other or collaboratively move together sometimes. We also said that we would need a method called rigging, so let's give an example to understand why rigging is useful.

When you talk, whether you are sitting or walking, the muscles and bones that are responsible for the talking are generally not affected by or affecting the other parts of your body. However, when you are walking, your legs rotate around the hip bones, and the rest of the system triggers other natural actions, such as swinging your arms, moving your shoulders slightly forward and backward, and so on.

In both cases where you have a local or system-wide dependency, we eventually move some of the vertices that make up a model. Since moving so many vertices is a lot of work, we use a structure we place inside the model to tell the necessary vertices where to move. The process to create such a structure is called rigging. In a way, rigging mimics what bones and muscles do in real life.

In this section, we'll work on a simple rigging process and rig a low-poly snake. Through this process, you'll prepare the model for animation, but first, we'll get to know some of the essential components, as follows:

- **Armature**: An armature, in simple terms, is a set of bones, but a better definition might be a framework serving as a control structure – what materials are to textures, armatures are to bones. So, the same armature could have multiple bones. Furthermore, the rigging process could involve many armatures if the system that's animated requires so.
- **Bone**: This is the most essential part of a rigging system. Without bones, there would not be armatures, therefore nothing to animate. In real life, when your bones move outside of their zone of freedom, you feel pain, so your body keeps things intact. There are similar ways to restrict a bone's freedom digitally, so to speak, so it works in tandem with other bones.

We'll first look at how to rig a model. For this effort, we'll utilize one armature and many bones. After adding constraints to some of the bones, the rigging process will be complete. So, in the end, we will use our rig to animate the snake.

Rigging

Now that the theoretical stuff is out of the way, we can focus on the practical aspects, mainly how to set up armatures and bones. To focus on the rigging process, we'll use a low-poly snake model. The Snake.blend file in Chapter 5's Start folder is a good starting point, and by the end of this *Rigging* section, you'll have reached what you see in the Snake.Rigged.blend file.

Besides these two files, we'll mention other complementary files that show the interim phase. As always, you can find all of these files at the URL mentioned in the *Technical requirements* section.

After you open the Snake.blend file, let's add an armature by performing the following steps:

1. Press *3* on your numpad to switch to the **Right Orthographic** view.
2. Press *Shift+A*.
3. Select **Armature**.

You can also find the result of the preceding operations in the Snake.First Bone.blend file. If your keyboard doesn't have a numpad, then you can click on the **X** axis in the gizmo in the top-right corner of **3D Viewport** until you read **Right Orthographic** in the top-left corner. The following figure should help you see what we have done so far:

Figure 5.4 – Beware the snake! On second thought, it doesn't seem to have a mean bone in its body

We now have a new object type in our scene: an armature. You can see it in **Outliner** too with two green stick figures next to its title. Right now, we have one bone in the armature. So, bone and armature kind of mean the same thing at this point. Our goal, in rigging, will be to create and distribute a bunch of bones inside the snake's mesh. So, let's add more.

We seem to have a problem, though. That bone we added earlier looks like it's occluded by the snake's tail. So, if we keep adding more bones and laying them out so that they align with the snake's body, we won't be able to see what we are doing. Luckily, the solution is a couple of clicks away. While the armature is still selected, you can expand **Viewport Display** in the **Armature** settings in the **Properties** panel and turn on the **In Front** option. This will make sure the armature is always visible.

Missing out on a numpad

Numpad shortcuts are helpful and they will make your life easier, especially during modeling and rigging when you need to view your work from certain angles often on. The following website offers eight different ways to mimic a numpad: `https://essentialpicks.com/using-blender-with-no-numpad/`.

Meshes are composed of vertices, faces, and edges. Similarly, bones are made of three components: **root**, **body**, and **tip**. The tip can be the root of another bone and vice versa. Just as we can go into **Edit Mode** for a mesh to change its inner parts, we can do so with an armature. So, select the armature and press *Tab*.

You should be able to click on and select the root and tip separately. When you select the structure in between the joints, it'll automatically select the root and the tip since it's all connected. *Figure 5.5* shows only the tip selected.

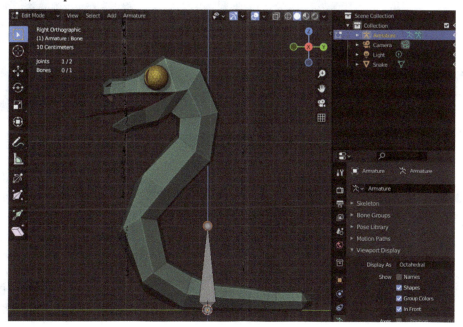

Figure 5.5 – The tip of the bone is selected in Edit Mode

> **Credit where credit is due**
>
> The snake model we are rigging in this section is an asset created by an artist known as Quaternius. You can follow his work at `https://quaternius.com`. We'll be using his other assets in later chapters as well. So, thank you for your generosity.

Now, we are ready to add more bones to the armature. We'll do that by first positioning that initial bone, then we'll add new bones coming off the tip. While still in **Edit Mode**, perform the following steps:

1. Select the root joint.
2. Press *G* and move the mouse so that the joint is somewhere in the middle of the snake's chest.
3. Click to finish grabbing.
4. Select the tip joint.
5. Press *G* and move the mouse so that the joint is somewhere near the Y axis but inside the tail.
6. Click to finish grabbing again.

A figure might be extremely helpful since all of this moving and positioning sounds a bit arbitrary. *Figure 5.6* is an example of what we have achieved in the last few steps.

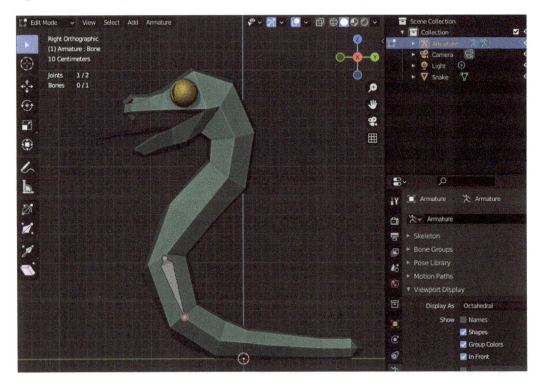

Figure 5.6 – A well-placed bone for our snake

Since up, down, or right concepts lose their meanings in the 3D space, it's important to have a simple yet effective way to represent the natural flow of bones. If you compare *Figure 5.5* and *Figure 5.6*, which correspond to the `Snake.First Bone.Editing.blend` and `Snake.First Bone.Position.blend` files, respectively, you'll notice that the structure between the joints is going in different directions. The broader part of the bone is closer to the root, and the narrower end of the bone is approaching its tip. For example, imagine your kneecap as the root and your ankle as the tip of one bone. Moreover, hip bone to kneecap, elbow to wrist, and so on.

We have to add a few more bones to our system. We'll do that by extruding the original bone. While still having the tip of the bone selected, perform the following steps:

1. Press *E* to start extrusion.
2. Move the mouse in the right and bottom direction so it follows the tail's form.
3. Click to finish extrusion.
4. Repeat *Steps 1 to 3* until you have four bones of roughly the same length.

The result is shown in *Figure 5.7*, and you can also open the `Snake.Tail Bones.blend` file to compare your result.

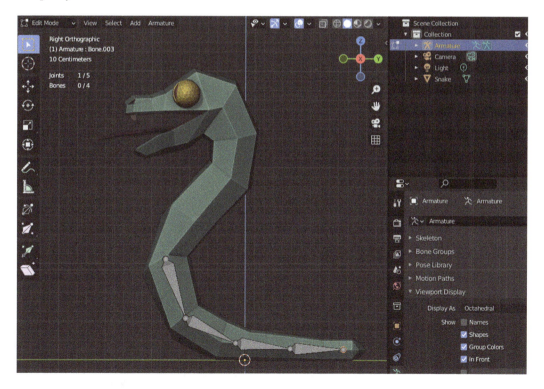

Figure 5.7 – Four bones that make up the tail

Importance of clicks

Similar to finishing a grabbing operation, extrusion needs a final click to solidify the position of an extruded object. Hence, throughout the rest of this chapter, when you follow a step where you see the word *extrude*, you are expected to click and finalize the extrusion when you are happy with the object's position. If you prematurely terminate the extrusion, you can always hit *G* and grab this new object to move elsewhere and continue extruding if you wish. Thus, click to finalize both grabbing and extrusion, and use these two handy methods as often as you need. Also, if you change your mind while extruding, right-clicking will cancel this operation.

Extrusion helped us do a few things at once. We have created a new bone, positioned it correctly so its root aligned with the previous bone's tip, parented this new bone to the previous bone, and finally, moved its tip to where we'd start the next bone.

We're halfway through adding bones to the snake. That being said, now is a good time for a bit of housekeeping. We'll be referencing some of these bones later, so it would be prudent of us to rename them now. If you have been paying attention to the new bones' names after the extrusion, you must have seen that they are labeled in a format that goes like **Bone.00X** where X is the succeeding bone's number. To rename all of the bones you have added so far, perform the following steps:

1. Select the original bone.
2. Press *F2* and rename it to `Tail.1`.
3. Repeat the preceding two steps for the rest of the bones so that their names look like **Tail.X**.

Let's move on to adding bones for the torso. For this, we are going to utilize the original bone, which is now renamed **Tail.1**. Some of the decisions that you'll make while rigging your models will depend on the situation you are going to use the rig for. It would have been perfectly possible to start the bones from the head and go all the way to the end of the tail. However, we know that this snake will have an inclination point, mainly where the torso and tail bones meet. Therefore, you need to perform the following steps:

1. Select the root of **Tail.1**.
2. Press *E* to extrude a new bone in the right and top direction, following the torso.
3. Repeat *Step 2* twice more so that you have three bones in the end.
4. Select each new bone and rename them to look like **Torso.X** where X is a consecutive number starting at 1.

The result is what you see in *Figure 5.8* and in the `Snake.Torso Bones.blend` file.

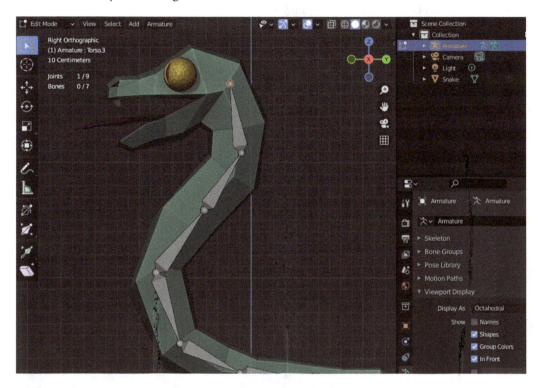

Figure 5.8 – New bones have been added following the torso to the head

We can now plan the remaining bones. We'll be concerned with only two bones for brevity's sake: the head and mouth bones. If you have been following all along, the tip of **Torso.3** should still be selected. If not, select it, then perform the following steps:

1. Press *E* to extrude a new bone to the end of the snake's nose.

2. Select **Torso.3**'s tip again.

3. Press *E* to extrude a new bone to the end of the snake's mouth.

In the end, the fully constructed skeleton, which you can find in the `Snake.Full Skeleton.blend` file, will look like what you see in *Figure 5.9*.

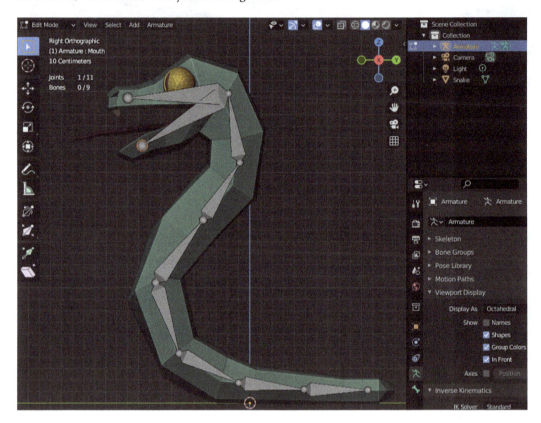

Figure 5.9 – The skeleton of our snake is complete

We are done with the skeleton. To complete the rigging, we need to add two more bones, which are usually called control bones. The following is an explanation of why a simple skeleton, although necessary, is still considered less than ideal. It has to do with the following two conflicting concepts:

- **Forward Kinematics (FK):** When you have a series of bones and you want to move the extremity bones, for example, a thumb in a human's hand, the motion would have to be calculated while considering all of the position and orientation values for all of the interim bones starting from the shoulder joint. Thus, the motion starts off at the root and goes forward.

- **Inverse Kinematics (IK):** This is a much more efficient method where, following the preceding example, by moving a thumb, all of the connected bones determine their state in reverse order one at a time, instead of calculating the overall system's behavior. Thus, the moving bone dictates how the bone behind should behave, and that bone behind does the same all the way to the root.

We prefer IK in our exercise since it's much more convenient to use, and it is widely accepted in the industry. If you would like to get more in-depth information, especially on the math aspect of FK and IK, refer to the following two pages:

- `https://www.sciencedirect.com/topics/engineering/forward-kinematics`

- `https://www.sciencedirect.com/topics/engineering/inverse-kinematics`

To introduce IK to some of our bones, we need to create control bones that will propagate the motion to the rest of the bones. Although these control bones will look like they are part of the skeleton visually, they will be decoupled from the skeleton. Right now, all of the bones that have been extruded have been automatically parented. So, we'll need to unparent our two control bones once we extrude them off the end bones.

It would seem one of these bones could be coming off the **Head** bone, and the other control bone, by symmetry, could be coming off the **Tail.4** bone. Assuming you are still in the **Right Orthographic** view, in order to create these bones, you need to perform the following steps:

1. Extrude a bone in the left direction off the tip of the **Head** bone.

2. Rename this new bone as `Head.IK`.

3. Extrude a bone in the right direction off the tip of the **Tail.4** bone.

4. Rename this new bone as `Tail.IK`.

We have created two new bones, but they are still attached to the skeleton. So, we need to separate them. *ALT+P* is a shortcut you can use to clear the parent relationship, but we'll do the decoupling somewhere else since we'll have to turn off another setting too. So, let's do both at the same time, as follows:

1. Select the **Head.IK** bone.

2. Turn on the **Bone Properties** tab (the green bone icon) in the **Properties** panel.

3. Expand the **Relations** section in that tab.

4. Clear the parent by clicking on **X** in the name field.

5. Turn off the **Deform** option.

6. Repeat *Steps 3 to 5* for the **Tail.IK** bone.

The `Snake.Full Skeleton.IK.blend` file contains all of the progress you have made so far, but let's explain what we have done in the last several steps. We used to see the **Armature** properties, so we asked the **Properties** panel to show another view to display bone properties. We broke the connection of our control bones with their parent. Since there is no parent, the **Connected** checkbox automatically switched itself off. Lastly, we turned off a setting that's the crux of all this whole operation: **Deform**.

If you recall what topology is and why we use a rigging system to animate systems that bend and stretch, then you'll know that deformation is the key. We want the skeleton of the snake to deform the mesh it's in. However, we wouldn't want that for the control bones since we'll use these to dictate the overall motion. So, they should not be deforming anything.

That being said, they will be responsible for IK, which is the last missing piece to the rigging. To complete the rigging, we need to add the **IK** ingredient, and we'll do that in **Pose Mode**.

In *Chapter 1*, *Creating Low-Poly Models*, we went back and forth between **Object Mode** and **Edit Mode**. In this chapter, we've been in **Edit Mode** all this time to move the parts of a bone and extrude new ones. Bones can be in another mode, **Pose Mode**, with which you can define the relationship of the bones with each other by introducing constraints. Consider this new mode as editing the behavior of the armature, hence how the model will *pose*.

Assuming you are in **Edit Mode** already, press *CTRL+Tab* then press *2* to switch. Or, if you are in **Object Mode**, then *CTRL+Tab* will take you directly to **Pose Mode**. Keep in mind that this works if you have a bone or the armature selected. Alternatively, the dropdown in the top-left corner can help you to be in the right mode. We're now ready to add **IK** constraints as follows:

1. Select the **Tail.4** bone.

2. Turn on the **Bone Constraints Properties** tab (the blue bone icon with a strap around it) in the **Properties** panel.

3. Choose the **Inverse Kinematics** option in the **Add Bone Constraint** dropdown.

4. Repeat *Step 3* for the **Head** bone.

We have added the missing **IK** component to two bones. Maybe you noticed that the constraint was not added to the control bones but to the bones just before them. We'll now map some of the **IK** constraints' values to use the control bones. To do that, while the **Head** bone is selected, perform the following steps:

1. Click on the square icon in the **Target** field of the **IK** constraint.

2. Select **Armature** in the options.

3. Click on the bone icon in the **Bone** field of the **IK** constraint.

4. Select **Head.IK** in the options.

This will designate **Head.IK** as the control bone for the **Head** bone. So, from now on, whenever you interact with **Head.IK**, it will control the **Head** bone that is connected to the other bones all the way to the root. That's why you see a dotted yellow line going from the tip to the joint in between the **Torso.1** and **Tail.1** bones.

Let's associate **Tail.4** and **Tail.IK** by following the preceding recipe so that interacting with **Tail.IK** can dictate the tail bones' behavior. Select **Tail.4** then perform the following steps:

1.   Select **Armature** in the options after clicking on the square icon in the **Target** field.

2.   Select **Tail.IK** in the options after clicking on the bone icon in the **Bone** field.

3.   Change the **Chain Length** value to **3**.

The first two steps in the preceding set of instructions are pretty much exactly the same except that we picked the appropriate bone. The last step introduced a new concept that tells the control bone how far down the chain of bones the root bone is. The dotted line moved accordingly. The final result is what you see in *Figure 5.10*.

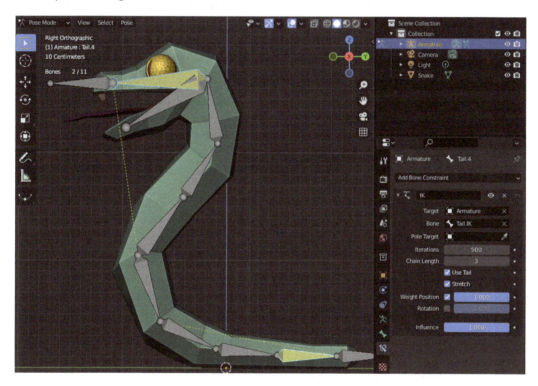

Figure 5.10 – A fully rigged snake

We've been doing all of this work so that the armature would be part of the snake. However, if you look at **Outliner**, you can still see that these two objects are separate. It's time to really connect the skeleton to the snake's mesh as follows:

1.   Switch to **Object Mode**.

2.   First select the **Snake** mesh, then **Armature** by holding down the *Shift* key.

3.   Press *CTRL+P* to bring up the **Set Parent To** menu.

4.   Choose **With Automatic Weights**.

When you parent the armature to the mesh, two things will happen. First, **Snake** in **Outliner** will be moved as a child under the **Armature** item. Second, **Snake** will be assigned an **Armature** modifier that will build the connection between these two objects.

In the end, the armature will designate its bones to nearby vertices so that when a bone moves, it mobilizes the associated vertices. It's as if some vertices that are closer to a particular bone *weigh* more in terms of priority. Thus, you won't see a tail bone move far away vertices that much.

Phew, the rigging is finally complete. As you may have noticed, all of this creating and separating bones, adding constraints, adjusting settings, and so on could sometimes become a tricky business. You get visual clues as to which bone is doing what and how they are connected, but the scene could quickly get cluttered with gizmos. Like anything else, though, you get used to doing it with practice. On that note, you'll find links to more advanced rigging material in the *Further reading* section.

We have provided the `Snake.Rigged.blend` file both in the `Start` and `Finish` folders for you to compare your results. You can also use this file as a starting point in the following section. Since we deemed that rigging was necessary for animation and that our rig is done, we can now turn to a new section where we'll get to know the **Animation** workspace of Blender.

Animating

We're about to animate our snake. We've prepared a skeleton and introduced two control bones to construct a rig. In this section, we'll use this setup to create an attack animation. Using the methods presented in this section, you can create different animations for your models and store these animations with the model in the same file.

Let's switch to the **Animation** workspace to take advantage of a more suitable set of interfaces. The layout will change to mainly two side-by-side **3D Viewport** panels and what looks like a timeline underneath. There are actually two panels at the bottom, as follows:

- **Dope Sheet**: We'll work with keyframes soon to mark the defining points when parts of your model move over time. For example, a frog can have one keyframe for its resting position, then another keyframe defined as its highest jump level later in time.

- **Timeline**: This is a simpler version of **Dope Sheet**. It is represented with a clock icon and lets you see things at a higher level. We won't utilize this interface that much, but it's useful to set the **Start** and **End** keyframes of your animations.

Besides these two editors, there is also the **Graph Editor**, which you can access by clicking the icon in the drop-down menu in the top-left corner of any panel. Actually, let's do that by changing the left **3D Viewport** into a **Graph Editor**. When you are done, you should see something like the following:

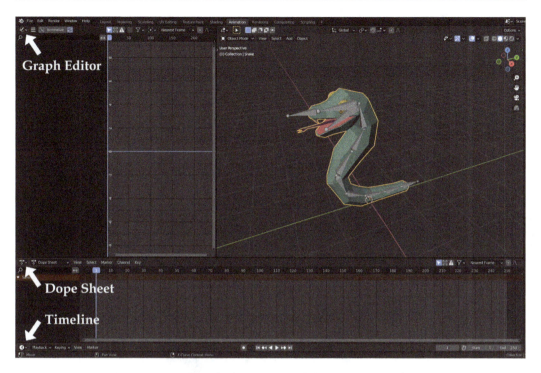

Figure 5.11 – We have further customized the Animation workspace

We have everything we need to animate the snake. We'll start with an attack animation. For this, we'll move the head forward and raise the tail to depict a menacing pose. Start by switching the 3D perspective to **Right Orthographic** by pressing *3* on the numpad and performing the following steps:

1. Go into **Pose Mode**.
2. Select the **Head.IK** bone.
3. Press *I* to insert a keyframe and select **Location** in the options.

This operation will add a key to the first frame in **Dope Sheet** as well as populating some elements both in **Dope Sheet** and **Graph Editor**. So far, so good. Take a look at what's added to the animation editors and expand the **Head.IK** title in both editors to see what exactly is happening under the hood. We are marking the location of the **Head.IK** bone.

For the next event in the snake's attack animation, we need to move the snake's head forward and key (mark) its new location. For this, we need to select a new frame in the timeline as follows:

1. Change the frame value from **1** to **10** (just to the left of the **Start** section in **Timeline**).
2. Press *G* and move the head slightly to the left and up.
3. Press *I* to insert a keyframe and choose **Location** again.

This should add more elements – more specifically, curved lines – to **Graph Editor**. This is good because you can use those curves to fine-tune how the action will start and end—more abruptly or smoothly, which can be used for more dramatic effects. We leave it to your artistic interpretation. What we can do, for now, is finish the head's motion so that it goes back to its original position, as follows:

1. Change the frame value from **10** to **25**.

2. Press *Alt+G* to reset its position to the original values.

3. Press *I* to insert a keyframe and choose **Location** again.

Figure 5.12 shows our progress so far.

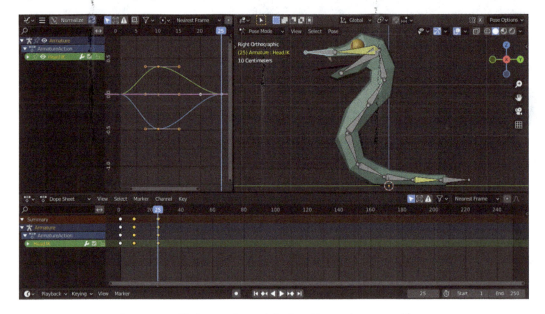

Figure 5.12 – We have animated the head bone via a control bone

In the end, we have moved the torso bones by animating the **Head.IK** bone. That's why we have implemented a control bone instead of moving the individual torso bones. Additionally, we haven't done anything special to the **Mouth** bone, but that's also moving to keep up with the head.

Let's do something similar with the tail, as follows:

1. Set the frame to **1**.

2. Select the **Tail.IK** bone.

3. Press *I* to insert a keyframe and choose **Location**.

4. Set the frame to **10**.

5. Press *G* and move the tail slightly to the top and left.

6. Press *I* to insert a keyframe and choose **Location** again.

7. Set the frame to **25**.

8. Press *Alt+G* to reset the position.

9. Press *I* to insert a keyframe and choose **Location** again.

In this pose, the tail naturally looks angry, which accentuates the head's motion. By the way, where is your head? If you look in **Dope Sheet**, the keyframes for the head animation are gone. Blender only displays the keyframes for the selected object to keep the interface clean and simple. You can display everything by toggling off the **Only Show Selected** button, which looks like a select icon in the header of **Dope Sheet**. There is a similar button in **Graph Editor**; if you disable both, you should have something similar to what you see in *Figure 5.13*.

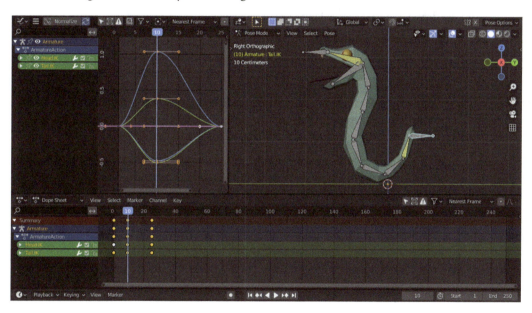

Figure 5.13 – Both the head and tail keyframes are visible in the editors

You can also refer to the `Snake.Animated.blend` file in the `Finish` folder.

We've completed our first animation. If you would like to create another animation, where would you do it? It seems that we could keep adding more keyframes to the timeline. However, how would we know which keyframes are responsible for a particular animation?

We can answer this question both in Blender and Godot contexts. Actually, once we understand how to create separate animations in Blender for the same model, we'll have practically prepared our animations to be shipped to Godot. To do this, we'll discover **Action Editor** in the following section.

Getting animations ready for Godot

Creating separate Blender files for different animations would be extremely unwieldy. If only we had a way to store multiple animations in the same file. Luckily, there is. We need to use a new interface called **Action Editor** for that. Let's see how we can use it to create another action for the snake.

There is a dropdown in the top-left corner of the **Dope Sheet** panel. Although that whole panel could be considered as the **Dope Sheet** panel, we have been using its default view. This is similar to how **3D Viewport** works. When we were switching between **Object Mode** and **Edit Mode**, we were still working in the same **3D Viewport** panel but in one of its specialized views. In other words, these dropdowns customize the panel you are in. To switch the **Dope Sheet** panel to its **Action Editor** view, perform the following steps:

1. Expand the dropdown that shows **Dope Sheet**.
2. Select **Action Editor** in the options.

This will reveal the title of our first animation, **ArmatureAction**, in the middle portion of the **Action Editor** header. This is a lackluster action name. The snake deserves better. Let's change it by clicking its text and typing `Attack`. Now, you have just changed the default name to something you can easily keep track of. Moreover, when we import this model into Godot and we want to trigger the correct animation sequence, we'll use this action name. Let's create more actions as follows:

1. Click on the second icon next to the action title (the icon with stacked papers).
2. Change this new action's title to `Idle`.

This will actually create a copy of the first animation. Except for its title, everything is the same, but we can now change the features of the animation that match the title we just gave. In most games, the idle state of characters usually looks calm, but they have a slight bobbing up and down motion that indicates the character is alive but otherwise in a neutral state. Our idle action involves performing the following steps:

1. Set the frame to **10**.
2. Select the **Head.IK** bone and reset its position by pressing *Alt+G*.
3. Press *G* and move the bone ever so slightly downward.
4. Press *I* and choose **Location**.
5. Repeat *Steps 2 to 5*, but move the **Tail.IK** bone slightly upward.

Let's do one more thing and test our new action. Change the **End** value in **Timeline** to **25** and click the play button. This will let you see the action in a looped manner so that you get a sense of whether the locations in the animation are good enough. Make more corrections to the location of the head and tail control bones if you would like, but remember to set their values by pressing *I*.

Our snake is idling, up and down, perhaps waiting for a target to attack. By using the dropdown to the left of the action's title, you can switch between different actions.

Congratulations! You have officially created two animations. If, at times, it was difficult to follow the instructions, you can find a fully finished example in the `Finish` folder in the `Snake.blend` file for further studying.

We have done a lot in this chapter. It's time to summarize our efforts.

Summary

This chapter started off with a discussion about which software (Blender versus Godot) would be suitable for animations. We exemplified different cases of animation and determined that Blender is the right choice for animating systems that have individually moving parts.

We then discussed the importance of good geometry, better known as topology, since not everything that looks good is good enough from an animation perspective. Once the system is in motion, the vertices, faces, and edges will act like a wrapper around a skeleton. If you know you'll be animating your model, you might be careful in how you create the geometry better ahead of time.

Nevertheless, if such an early option is not always possible, to prevent tearing and creasing that might occur in certain areas of a model, we introduced the grab option. It can help you resolve problematic parts by moving them to a different location.

As soon as the distribution of vertices is in a favorable place, then the rigging can start. This is, in fact, one of the most advanced topics for most artists who are learning any 3D modeling software. It helps sometimes to think of rigging as a bunch of strings that control a puppet. Like a puppet master, you need to know which string controls which parts. To that end, we introduced IK, which has advantages over a more direct, also known as FK, approach.

After we created a rig for a snake, we discovered the animation workspace. Since the rigging depended on control bones via IK, our animation was done effortlessly. Along the way, we learned how to move parts of a rig and keyframe their properties. In our simple case, it was only location, and we kept the motion on one axis.

Lastly, we got to know how we could store two animations, rather actions, for the same model. Once you have properly labeled actions, not only will it be easier for you to find them in Blender in the future, but you will also see the benefit of this practice later in Godot chapters.

You have completed five chapters that took you from creating models to adding animations to your models. Along the way, you've also learned how to construct and apply materials and textures. In the following chapter, we'll investigate how to export our work from Blender.

Further reading

We mentioned the importance of topology, and it could be challenging to know what constitutes good or bad topology. So, to see more examples and benefit from other people's expertise, refer to the following links:

- `https://blender.stackexchange.com/questions/140963/do-i-have-bad-topology`

- `https://www.reddit.com/r/blenderhelp/comments/speyjs/is_this_bad_topology/`

- `https://www.pluralsight.com/blog/film-games/ngons-triangles-bad`

Some 3D practitioners specialize only in animation. Although it's possible to animate some Blender objects without rigging them, for example, cameras and lights to move them around the scene, most online courses usually cover rigging and animation topics together. The following is a list of online courses and material for you to further your knowledge in both of these domains:

- CG Cookie: `https://cgcookie.com/courses?sort_category=140,179`

- Udemy:

  - `https://www.udemy.com/course/rigging-fundamentals-blender/`

  - `https://www.udemy.com/course/rigging-and-animating-low-poly-fps-arms-in-blender/`

  - `https://www.udemy.com/course/learn-3d-modelling-rigging/`

  - `https://www.udemy.com/course/blendercharacters/`

Additionally, while you are browsing for more training content, you might come across a topic called **Weight Painting**, which is helpful in determining how the rigging will prioritize the nearby vertices. We left it out for brevity's sake, but it's a topic you'll most likely want to cover if you want to be more thorough.

In the following chapter, we'll be slowly transitioning from Blender to Godot. So, this chapter was really the last hands-on Blender chapter. If you want to know more about what Blender can do, there are some really useful resources out there, in both written and video formats, offered by Packt Publishing, such as the following resources:

- *Blender 3D By Example* by Oscar Baechler and Xury Greer

- *Blender 3D Modeling and Animation: Build 20+ 3D Projects in Blender* by Raja Biswas

- *The Secrets to Photorealism: The PBR/Blender 2.8 Workflow* by Daniel Krafft

Part 2: Asset Management

In this transitional part, you'll learn how to move from Blender to Godot. An essential part of this workflow will be knowing which settings matter. By getting to know potential pitfalls and how to apply workarounds, you can prepare yourself for scenarios where you have to use third-party assets.

In this part, we cover the following chapters:

- *Chapter 6, Exporting Blender Assets*
- *Chapter 7, Importing Blender Assets into Godot*
- *Chapter 8, Adding Sound Assets*

6

Exporting Blender Assets

Your journey in Blender has taken you to this point, where you want to take your creations in Blender and deploy them in Godot Engine. We'll cover importing these assets into Godot in the next chapter, but first, we must make sure everything we have in Blender is up to Godot's standard. So, we've got to iron out a few kinks before exporting.

First, we are going to make sure the geometry of our models is fine. We have already talked about polygons; we'll dive deeper to understand them better to achieve models with better geometry. Origin points are an important concept in both Blender and Godot. We'll discuss why they are important and learn how to alter the origin points.

We have not discussed the dimensions of our models so far. However, more important than the dimensions of your models, we'll investigate a concept called **scale** or **scale factor**, which is crucial when you send your assets to not only Godot Engine but also to other game engines. The final part of getting your models ready is an organizational practice: naming your assets.

After we finish making our preparations, we'll need to convert our assets into a format Godot understands. To that end, we'll explore **glTF** and compare this format to a few others. Once Godot imports this file type, it will understand how to make sense of vertices, materials, and animations stored in a Blender file. We'll look into importing in the next chapter, though.

Lastly, just because we can transfer assets out of a Blender file doesn't mean we should be all-inclusive. We'll discuss which objects in a Blender scene are useful from a game development perspective. During this exercise, we'll also learn how to store our preferences for selecting the objects we want to export under **presets** so that we don't have to remember the export conditions every single time.

In this chapter, we will cover the following topics:

- Getting ready to export

- Exploring glTF and other export formats

- Deciding what to export

By the end of this chapter, you'll know what to do to get your models ready for export, choose an appropriate export format and configure it, and learn how to export only the stuff you want.

Technical requirements

This is a chapter about understanding some concepts rather than practicing, so you'll do a minimum amount of work, such as looking at the value of certain things and occasionally rotating some objects. You'll likely revisit this chapter later to remember how to export your work samples. So, it's OK to do a preliminary reading first and come back again for another read.

Wherever it's relevant in this chapter, the appropriate filenames in the `Start` and `Finish` folders will be mentioned. The files that contain the necessary assets have been provided for you in this book's GitHub repository: `https://github.com/PacktPublishing/Game-Development-with-Blender-and-Godot`.

Getting ready to export

There are plans to make the transition between Blender and Godot Engine more seamless in future versions. For example, you'll be able to deploy your Blender file directly in a Godot project and start accessing the elements from your Blender scene directly in Godot. However, we are not there yet, so we need to do a bit of housekeeping before we send our stuff to Godot.

The following is not a complete list, but it covers the most common problems many artists face when they go between Blender and Godot:

- Deciding what to do with n-gons
- Setting origin points
- Applying rotation and scale
- Naming things properly

Now, let's discuss these topics (problems) and their solutions. We'll start with more labor-intensive topics and finish off with easier things to take care of before you hit the export button.

Deciding what to do with n-gons

Let's give a formal definition of an **n-gon** and move on to its relevance in our work. Mathematically, a closed plane with n edges is an n-gon, but we use friendlier names for some of these n-gons. For example, a triangle is another name for a 3-gon. Moreover, for any number of edges equal to or more than five, we generally use Greek prefixes to describe them – this includes pentagons, hexagons, heptagons, and others. Lastly, a question for you to ponder on: what do you call a 4-gon, a square or a rectangle?

Although nothing is stopping you from creating 3D objects with faces that can make up any type of n-gon, you should avoid it in some circumstances. It's not a hard rule but it's something to keep in mind. So, why is this important for us?

We briefly discussed the role of a **Graphics Processing Unit (GPU)** in *Chapter 1, Creating Low-Poly Models*. Just as a reminder, a GPU takes a polygon and dissects it into the tiniest n-gon, namely a triangle. So, when you throw a bunch of complex polygons such as a pentagon or worse at the GPU, it processes these complex shapes to the best of its capability into triangles. This process is called **triangulation**. The following figure shows a few examples of triangulation:

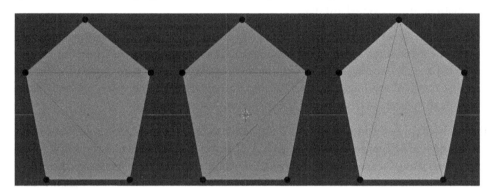

Figure 6.1 – The triangulation result could be different for the same polygon

Thus, when you leave the triangulation task to the GPU, it makes assumptions about which vertices should connect. Keep in mind that we don't want all the vertices to be connected, just the minimum number without creating any overlapping edges. So, for a pentagon, we can have five different triangulation cases. That's a lot of guesswork for a GPU to know which one you'd prefer.

In *Chapter 5, Setting Up Animation and Rigging*, we discussed the role of **topology**, which mainly involves distributing edges and faces. If you studied the content in more detail by following the URLs provided in that chapter, you must have come across the notion of edge flow. If you have a rig that's supposed to bend the model, you'll want the edges to follow a line as straight as possible into the bent part. Consequently, it pays off to do your own triangulation to create a smooth edge flow or simply avoid any n-gons altogether.

N-gons usually occur when you do loop cuts, but you can also create them accidentally while editing other parts of your model without noticing it. A quick way to get rid of them, if you can't avoid creating them, is to connect some of the vertices manually. You'll find an object with five vertices, hence five edges sharing one face, inside the Ngons.blend file in the Start folder. That's a 5-gon or a pentagon right there. Let's see how we can fix it:

1. Select the vertex at the top and one of the bottom vertices by holding *Shift*.
2. Press *J* to trigger the **Connect Vertex Path** operation.

This may not look much different, but you have added one more face by connecting those two vertices. You must have two faces now. Let's do something similar but pay attention to the number of faces shown on the right-hand side of the status bar. It should show **Faces: 0/3** after you do the following:

1. Select the vertex at the top and then the other bottom vertex by holding *Shift*.

2. Press *J* to connect these two vertices.

After your previous edits, your pentagon will look like the third case in *Figure 6.1*. If you fancy it, you can undo your steps and connect another set of vertices. Which vertices you should connect depends on your situation, so there is no hard rule.

Despite the number of vertices staying the same, you now have two more faces and two more edges compared to the initial state. Speaking of the initial conditions, take a look at **Tris** in the status bar, and reopen Ngons.blend without saving; you'll see that **Tris** in the status bar will still show **3**. That's because the GPU was implicitly triangulating the pentagon. You have now explicitly defined which vertices should connect, hence where the edges and faces should be.

Now that we have covered why and when it is important to fix the n-gons, here is a situation where you may not need to be concerned about n-gons at all. If you have a model that you know, for sure, you won't be animating (hence there is no rigging that would require a clean topology), then you can do without fixing your n-gons. Professionals insist on fixing n-gons because chances are the models will be animated, so they do it just in case. However, you now know you also have a choice.

Setting origin points

An origin point is a point where all your transformations start. This often sounds a bit technical, so sometimes, it's easier to think of it as the center of gravity. However, that might be a misleading definition because you can change the origin point for your models, whereas the center of gravity doesn't normally change in real life.

We must open Origins.blend in the Start folder to get to the bottom of origin points. For now, let's just look at the following screenshot:

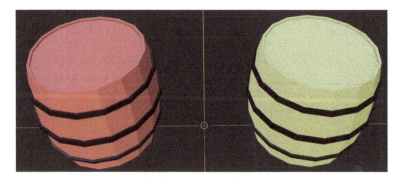

Figure 6.2 – These two barrels look very similar, but are they?

The Origins.blend file will contain two barrels, one painted in red and another painted in yellow. If you select the red and yellow barrels back and forth, you'll notice that an orange dot inside the outlined shape is in a different spot for each barrel. To get a better view of what's going on, you can switch to the **Right Orthographic** view by pressing 3 and observing that orange dot after you select either barrel. That dot is the origin point.

Follow these steps to understand the role of the origin point:

1. Select the red barrel.

2. Press *R* to rotate and then *X* to constrain the rotation axis. Then, type -45.

3. Select the yellow barrel.

4. Press *R* to rotate and then *X* to constrain the rotation axis. Then, type 45.

The values for the rotation were carefully selected to make these barrels tilt toward each other so that you can compare their final conditions. Although both barrels rotated the same amount, the yellow barrel seems to have leaned closer to the ground. To compare your results, you can refer to Origins-1.blend in the Finish folder, or take a look at the following screenshot:

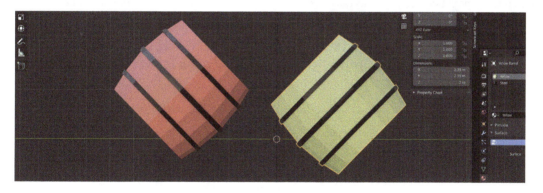

Figure 6.3 – Barrels rotated toward each other by the same amount around their origin point

Did you realize that both barrels were rotating around their origin point? We could take this a step further and place the origin point at the bottom of one of the planks of the barrel's body.

To make the barrel look like it's leaning around a more accurate pivot point, follow these steps:

1. Select the yellow barrel and press *Alt + R* to reset the rotation.

2. Go to **Edit Mode** and select the left-most vertex. Alternatively, hold down the middle mouse button to get a better view of the vertex that goes along the green **Y** axis.

We still need to complete a few more steps to set the new origin, but the following screenshot should help you find this mysterious vertex:

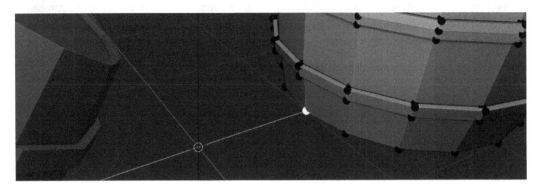

Figure 6.4 – This vertex will be the new origin point soon

In *Chapter 1*, *Creating Low-Poly Models*, we briefly mentioned 3D cursors. You might be used to working with other types of cursors, such as the ones you often see in a word processor or code editor. They usually blink regularly and place the character right there when you type on the keyboard.

Well, this is a 3D cursor, and it doesn't blink, but its role is similar. You can see it sitting where the **X** and **Y** axes meet in the preceding screenshot. To move that 3D cursor to the selected vertex and set a new origin, do the following:

1. Press *Shift* + *S*. A radial menu will appear and offer many choices for snapping.
2. Select **Cursor to Selected** or press *2*.

The choice we selected snapped the 3D cursor to the vertex you have selected. We are not quite done with moving the origin yet since we haven't told the *barrel object* where the new origin is. For that, we need to do the following:

1. Go back to **Object Mode**.
2. Right-click and choose **Origin to 3D Cursor** under **Set Origin**.

This will move the origin point of the barrel to the 3D cursor. That's why we had to move the 3D cursor to a specific vertex so that we could designate it as a new origin. The following screenshot shows the context menu and where to find the origin options:

Figure 6.5 – Setting the origin is a common operation, so it's part of the context menu

You can open `Origins-2.blend` in the `Finish` folder to see the yellow barrel applied with the same rotation from before, but, this time, the rotation is happening around a different origin point.

In the end, in most situations, setting a new origin point involves going into **Edit Mode** to select where you'll move the origin, then shifting the 3D cursor to this point temporarily so that you can set the origin in **Object Mode**. You could, of course, designate a completely arbitrary point outside the volume of your objects as their origin too.

An origin point will be used in Godot later, similar to Blender. If you set the origin point for a door at one of the hinges in Blender, rotating that door in Godot around the **Y** axis will use the hinge to revolve the door so that everything will look correctly calculated and adjusted.

Applying rotation and scale

This is, by far, one of the most important topics to take care of before you export your Blender asset. It has been mentioned several times in this book that looks can be deceiving. Applying rotation and scale falls under the false looks category. Let's understand this issue better by opening `Scale.blend` in the `Start` folder.

You should see two cubes, as shown in the following screenshot, that are on either side of the **X** axis. Also, the **Transform** panel is already expanded for you to look at the transform these cubes have, and you can use the *N* shortcut to toggle it on and off in the future. An object's transform is defined by its location, rotation, scale, and dimensions, but we're only interested in rotation and scale.

Those two cubes sure look the same, except one is green and the other is red, but they also are different in another way. Start by selecting the red cube, then the green cube. Do this a bunch of times while paying attention to what's changing in the **Transform** panel.

The following screenshot also shows you where you can find this panel:

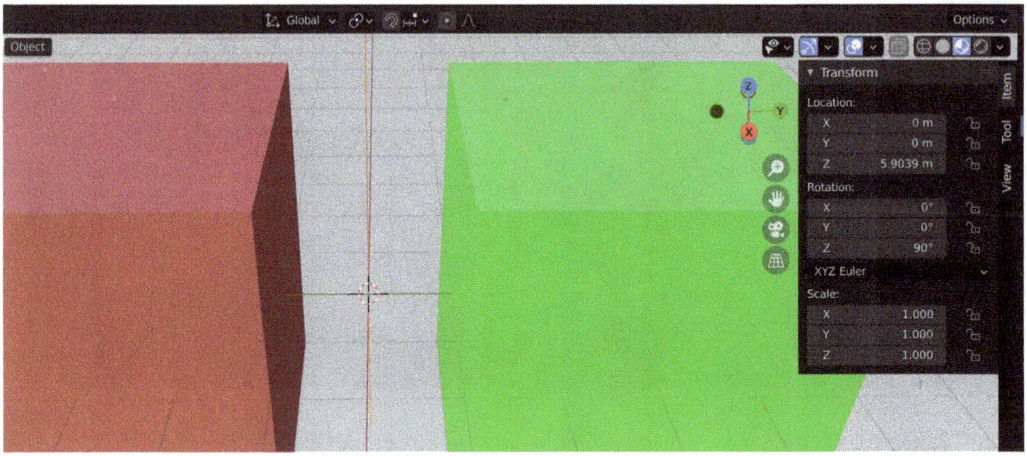

Figure 6.6 – The Transform panel is in the top-right corner of the 3D Viewport area

Both cubes' dimensions are 4 x 4 x 4 meters. Their location, individually, indicates where they are supposed to be. So far, so good. The scale and rotation values tell us a different story, though. So, how did this happen? Simply, the author of this file did what even the most advanced users sometimes do: they started modifying the properties of the red cube in **Object Mode**, whereas the green cube received its changes in **Edit Mode**.

Making such a simple mistake is quite common, and in fact, it may not even be considered a mistake because sometimes, you just want to select things and start editing without worrying too much about which mode the object is in. However, once you are done, you need to reset the rotation and scale back to 1 for game engines to do their job. This is one of the most common things people fix before they deploy their models to any game engine, so the situation is export format-agnostic. So, if you want to export your files as FBX so that you can import them into Unity, you'll still need to do this.

Luckily, the fix is simple. You can select the object that has a transform you want to fix, then press *Ctrl + A*. A popup menu will ask you what properties you would like to apply, which will reset the object's transform for the selected property. The fifth option, **Rotation & Scale**, is what we are looking for. When you trigger that option, you'll see that the red cube's rotation and scale values will reset to their default values.

After you import your models into Godot Engine, or another game engine for that matter, when your models behave in a weird way, such as some faces are missing or the animations are acting up, often, the rotation and scale are the culprits. So, make sure they are zeroed in before you export.

Naming things properly

Phil Karlton, who worked at Netscape, now a disbanded company that paved the way for browsing the internet with their web browser *Netscape Navigator*, famously uttered the following words:

"There are only two hard things in computer science: cache invalidation and naming things."

This quote is often passed around as a joke but, like most jokes, there is a hint of truth. If not in cache invalidation, there certainly is for naming things. Seeing meaningful names will make it easier for the future you or for a colleague to remember and understand what was done before.

When you start with primitive objects, Blender will label them for what they are: cube, plane, light, and so on. Your models will eventually get more complex at some point, and they will most likely have parts that will no longer look like a cube. So, keeping the original names will make your life harder at some point, both while working in Blender and Godot and even in another application if you use your exported assets.

So, give your objects names!

Wrapping up

You'll likely do some of these fixes more regularly than others. It's easy to forget to apply transformations, for instance, but it's an easy fix. Changing the origin point is a useful method during the modeling process for you to scale and rotate things smartly. In the end, you'll most likely leave it at its last position, so it's OK to come back to Blender to set it to its permanent position for your game to apply correct transformations later. Peruse the list of topics presented in this section as often as you need, and you'll develop a habit over time.

If you would like to practice the notions presented so far, we have prepared a `Fix-Me.blend` file in the `Start` folder. We wanted to design a simple heavyweight very fast, so that effort left the object with its default name. Also, its rotation and scale values look premature. While you are at it, you can also fix the n-gon and move the origin point to a different corner.

At some point, you'll eventually want to transfer your files to Godot. To that end, we often use exchange formats when both applications don't share a common file format. That'll be the case for us since we can't directly open and process Blender files in Godot. Therefore, we will discover a file format, glTF, that's been gaining popularity in recent years. It will help us transfer our work in Blender to Godot Engine.

Exploring glTF and other export formats

Compatibility between different software has always been a delicate matter. Actually, with most physical things, it is still a common problem even in modern life. Electric plugs and sockets, for example, come in different shapes and sizes in many countries. At the time of writing, 15 plug types are used worldwide according to https://www.worldstandards.eu/electricity/plugs-and-sockets/. You may want to make sure your devices are compatible before you leave home for a long distance.

It seems there is no consensus on what type of plug is best. Similarly, when it comes to exchanging data between different pieces of software, there are a plethora of options you could choose from. So, in the next few sections, we will discuss different types of export formats to see why we should choose glTF over other formats and how glTF is the better choice. Then, we will discuss glTF in detail.

Comparing glTF with other formats

Out of the dozen file formats Blender employs in its arsenal of export options, we'll focus on glTF because it works well with Godot Engine. That being said, let's present a few popularly used formats such as **Collada**, **FBX**, and **OBJ** first before we get to the good stuff:

- **Collada**: This format, which has DAE as its file extension, was conceived to be a data exchange format between 3D applications. This sounds promising at first, but although a game engine could be considered a 3D application, it's not – at least regarding the way this format was intended to be used. Collada was designed more for exchanging information between more classic 3D authoring programs such as Blender, Studio Max, Maya, and others, but not so much for game engines.

  It's based on XML, so you can open a Collada file with a text editor. This format fell out of favor over time since the specifications were ambiguous and have been incorrectly interpreted and implemented. For earlier versions of Godot, especially before glTF was out, Collada used to be the preferred file type. Now, we have glTF as a much better option.

- **FBX**: This is a proprietary file format offered by Autodesk. Since there are no official format specifications available to the public, and FBX's license doesn't let open source projects use FBX, even if the specifications are privately acquired, there have been attempts to reverse-engineer this format to write exporters for it. That's how Blender implemented the FBX exporter to the best of their guesses.

  Additionally, Godot engineers did their best to implement an FBX importer. Nevertheless, all this has been a bit of guesswork since the specifications are not open. To prevent hidden surprises and for a more seamless transition over to Godot, we won't use this format.

- **OBJ**: This is a simple plain text data format created by Wavefront Technologies. So, yes, this too can be opened with a text editor. Plain text data formats offer ease of editing, but since they are not compressed files, it's often slow to parse and import them. OBJ suffers from a different problem, though. It can't store animations and light sources, but it's a simple and good format to primarily hold mesh information.

  This also means it doesn't store material and texture information. To achieve that, you need to create an MTL file alongside the OBJ file you are creating. OBJ is an old and reliable format and is considered an industry standard, but it's not cut out for modern game engines.

Now that we have seen which formats we won't use, let's focus on what makes glTF a better choice for us. We'll do this by providing a brief history of glTF, followed by presenting which settings we must choose in Blender's export settings for our efforts.

Introducing glTF

Short for **Graphics Language Transmission Format**, glTF was first released in 2015 by Khronos Group, a member-driven non-profit consortium founded and empowered by many big corporations. Not every member corporation is in the digital content creation business, but they have a stake in the consortium because Khronos maintains other standards such as OpenGL and WebGL, two well-known graphics APIs that serve many industries.

The discussion about the reliability of a file format might be important at this point, especially if you are planning to reduce long-term maintenance problems and costs. For example, how many of us remember the early internet days' video file formats? Just to name a few, there was RealMedia, QuickTime, DivX, and many others, for which we'd have to install codecs, plugins, and more just to watch a few cat videos. Our desire to watch our furry companions never changed, thankfully.

Nevertheless, things coalesce eventually, and it gives way to better and more maintainable file formats. Hence, guidance from a standards group such as Khronos is a good thing since they ensure that the file format receives proper attention and stays up to date with the ever-changing needs of the industry. glTF is one of these healthy cases, and the fact that it's open source and many corporations would like to support it is a good sign. It would be a terrible day if you had a bunch of assets sitting in your game engine one day and you learned that you can no longer export in that file type. What would you do with the existing assets – throw them out and convert them into a new format?

Now that we've had a brief history lesson, let's get to know the relevant parts for us. We'll utilize Blender's glTF implementation, which supports the following features:

- Meshes
- Materials (Principled BSDF) and Shadeless (Unlit)
- Textures
- Cameras
- Punctual lights (point, spot, and directional)
- Extensions
- Custom properties
- Animation (keyframe, shape key, and skinning)

We won't use even half of this feature set. We discussed why we won't fuss over cameras and lights in *Chapter 4, Adjusting Cameras and Lights*, for we'll set them up when we are building our game in Godot.

A quick note on what Blender's glTF exporter does with meshes: n-gons will automatically be triangulated. So, it won't be left to the GPU's mercy. The *Deciding what to do with n-gons* section of this chapter covered how to split faces into triangles if you need a reminder on how to triangulate manually.

Let's finish this section off by presenting three different flavors of glTF you can use. To access the list of variations, you've got to choose the **glTF 2.0 (.glb/.gltf)** option after expanding the **Export** menu item in the **File** menu. In the pop-up screen that appears, you'll see a **Format** dropdown on the right-hand side, which will show the variations that you can see in the following screenshot:

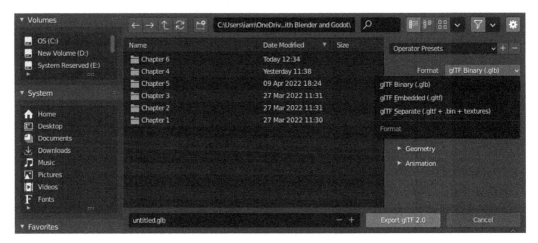

Figure 6.7 – Three possible variations you can use for a glTF export

These format variations will work the same, regardless of what settings you choose. We'll cover these in the next section, so first, let's get to know what each variation does:

- **glTF Binary**: This is the default option that will create a file with the `.glb` file extension. We'll use this variation throughout this book, and you'll most likely use it in your workflow as well since it stores everything you need in one file, and it's compressed. This makes it easy to share with other people and transfer over the internet.

- **glTF Embedded**: This is similar to the binary option, except it converts all the data into a JSON text format, similar to some of the other file formats we discussed earlier in this chapter. This will result in a file with the `.gltf` file extension and will make the file size larger but open to easy modifications with text editors if you wish. There is no practical reason why we should choose this variation over the binary option.

- **glTF Separate**: This last option will create many files: one file with the `.gltf` file extension, similar to the one you get if you choose the **Embedded** option, then a `.bin` file that holds the data, and optionally all the textures you've used with either `.jpg` or `.png` extensions. So, it likes to keep things separate. Since the data is stored in the `.bin` file, it keeps the `.gltf` portion smaller, unlike the **Embedded** variation. Nevertheless, there is still no practical reason for us to prefer this format. Also, if you had to send your model away, you'd have to remember to send all the separate parts too.

Regardless of the variation, the importing software will follow the glTF instructions set by the Khronos standards group to create your models, materials, animation, and others. So, choosing a variety may only be needed when it's necessary and for more advanced cases. For our work in this book, the binary variation will satisfy our needs.

Now that we know which variation is best for us, we must reflect on our own needs so that we can tick the right options in the exporter's interface. That's what we'll cover in the next section.

Deciding what to export

Not everything in your scene should be exported. For example, as mentioned previously, we will create the camera and light conditions for the game world inside Godot Engine. So, once that's done, there is no need to keep a camera and light object in your Blender scene. However, they might be useful for you to take test renders to get a better feeling for your scene without constantly exporting your models to Godot. In this section, we'll determine the better export candidates and how to use the export settings to facilitate that.

The export options are categorized, and we'll go through some of the options where appropriate. We'll do this by discussing how these options relate to the objects you have in your scene. Note that the export window is separate, so you don't need to close it before you select your objects in the scene. You can go back and forth between these two windows during this effort.

Include

Although the category's title is straightforward, the implications of what to include might be very important. By default, none of the options in this category are selected. So, it's up to your workflow. There are two groups you will see when you expand this section:

- **Limit to**: This is where you select what you want to include specifically as a mesh. We'll discuss this in more detail in the upcoming paragraphs.

- **Data**: Anything that is not a mesh could be considered data. For example, cameras and lights are not physical objects with mesh information but complementary tools that help you render a scene. We'll leave everything under here unchecked.

By default, all the options for both groups come unchecked. We've already said to leave the data untouched, but out of the four choices you can select under the **Limit to** section, the most important one is **Selected Objects**.

If you leave this unchecked, then Blender will include everything in your scene. This means that at the end of our exercise in the *Setting origin points* section, when we had two barrels, Blender would try to export both of those barrels. That's not something you'd most likely want. Chances are you'd want to design a barrel and export only that to Godot. So, we've got to have the **Selected Objects** export option checked first. Then, we need to go into our scene and select the object(s) we want to export. There might be some inconvenience in doing this so easily, though.

We have been designing relatively small models with a few different parts. The greatest number of separate parts we designed was with the three distinct parts of a barrel. In the future, during your work, if you happen to have a dozen or more parts in your Blender scene, it will quickly get tedious to select all these parts again and again before you hit the export button. If only we had an option that would not export the camera and light but what we deem as important so that we can have the best of both worlds…

That option is **Visible Objects**. Start by deselecting **Selected Objects** and keep the **Visible Objects** option on. For this option to work for us, we need to hide the camera and light objects so that they are no longer considered candidate objects to the exporter. You can do that by clicking the eye icon in the **Outliner** area for any object you don't want to export.

In the end, you have a mixed bag of solutions when it comes to what to include in your export. There are no right or wrong answers here, but you must choose what's efficient for you.

Transform

We'll cover this category for the sake of completeness. You'll rarely touch this category since it has one and only one option, which is on by default. Let's explain why, though, and learn what **+Y Up** means.

In Blender, the three axes or the coordinate system, **XYZ**, is set up, so the **Z** axis defines how tall or elevated an object is. In some other applications, such as Godot Engine, the **Y** axis is used as the going up axis. So, the higher the **Y** position of an object is in Godot Engine, the higher it sits in the game world. Therefore, this Blender export option converts Blender's **Z** axis into Godot's **Y** axis. It's a handy thing, so you don't have to arbitrarily rotate your models so that they match the correct direction.

Geometry

We'll leave most of the options under this category as-is and only discuss what matters to us. These options are as follows:

- **Apply Modifiers**: We first discovered modifiers in *Chapter 1, Creating Low-Poly Models*. We used a few that helped us model a barrel in no time. The fact that you can stack modifiers and change the order of operation is great. However, they are temporary additions to the core objects. So, unless you turn this option on in the export settings, the base object will be exported without any modifiers applied. This will make your objects look quite awkward and primitive in Godot.

- **Materials**: The default status of this option is to export all your materials. This might be a good thing for a beginner or quick results. When we discuss materials again when we cover Godot, and should you decide to make your own materials in Godot, you may want to pick the **No Export** choice so that they are no longer included in the resulting file.

In more advanced export scenarios, you may want to enable the **Loose Edges** and **Loose Points** options as well, so you can keep the loose geometry as a part of the exported file.

Animation

We won't change any of the default options in this category. We discussed how to create multiple animations in the *Getting animations ready for Godot* section of *Chapter 5, Setting Up Animation and Rigging*. The default settings will take care of converting the animations – more specifically, actions.

Creating presets

If you find yourself turning some of the options on and off under certain scenarios and memorizing the correct combination is becoming hard or monotonous, you can create a preset of export options. The top part of the export options has a dropdown with two buttons next to it. Using that area, you can create your own presets – perhaps one for a selected objects case and another one for a visible objects case.

Choosing the correct export options depends on the different conditions your project requires. So, you must experiment and find what works best for you. At some point, you'll import the result into Godot Engine to visualize the glTF file. However, that might be a lot of work to go in between two applications if you want to get a quick feeling about your creation. The following are two options you can use to preview glTF files:

- glTF Viewer at `https://gltf-viewer.donmccurdy.com/`
- Microsoft 3D Viewer

This concludes the investigation of the export options that are relevant to our case. Let's see what other discoveries you have made so far.

Summary

This chapter was mainly about making your work compatible with Godot Engine. To that end, we needed to go over a few different topics.

Firstly, we wanted to make sure our models had received the correct final touches. This involved getting rid of n-gons and converting these polygons into more manageable and ideal triangular faces. After that, you learned how to set origin points for your models, which may also be helpful during the modeling phase. Making transformations permanent is essential, so that's something to remember if your models, especially during animations, behave awkwardly. Then, we looked at the idea of naming things meaningfully. This is something you'll eventually find yourself needing more and more down the line when you have more experience.

Then, out of the many formats Blender offers for exporting assets, we evaluated a few, such as Collada, FBX, and OBJ. During that effort, we presented that glTF has become the de facto format for communicating between Blender and Godot. Lastly, we discovered some options for the glTF exporter and presented a few likely scenarios you may wish to employ. Finally, you learned how to store the export options that work best for you.

Now, we're ready to start importing our Blender assets into Godot. That's exactly what we'll do in the next chapter. In a real-life scenario, chances are you'll be conducting the operations presented in this and the next chapter quite often in almost every phase of your game development journey. Let's give you a few more useful resources before we move on.

Further reading

Khronos Group is a maintainer of many other standards we use day-to-day. This is thanks to their impressive list of members, which you can view at `https://www.khronos.org/members/list`.

We primarily used their glTF standard. The following links provide more technical information about it:

- `https://docs.fileformat.com/3d/gltf/`
- `https://docs.fileformat.com/3d/glb/`
- `https://www.marxentlabs.com/gltf-files/`
- `https://www.marxentlabs.com/glb-files/`

Thanks to its nifty specifications, the glTF exchange format has been gaining popularity not only in the gaming industry but in other industries as well. Here is NASA's famous *Voyager* spacecraft in all its glory: `https://solarsystem.nasa.gov/resources/2340/voyager-3d-model/`.

You may have come across websites where Collada is still used for Godot projects. Perhaps you already have access to a large repository of Collada files. If you would like to give it a try, but with a bit more finesse, here is a GitHub repository that can help you: `https://github.com/godotengine/collada-exporter`.

Last but not least, cleaning up your models and keeping them export-ready will be an ongoing task. Blender's user manual has a page on many tools and methods you can use to help you in your efforts: `https://docs.blender.org/manual/en/2.93/modeling/meshes/editing/mesh/cleanup.html`.

7
Importing Blender Assets into Godot

You've come a long way. Your models are ready. Their scale and rotation values are fixed. What's left to do? Import them into Godot, of course! Hopefully, you'll find the importing process much more straightforward. This is a transitional chapter that covers mostly Godot topics with a minimal amount of Blender involvement.

We'll start this chapter by showing you how to create game objects using your imported models with the click of a button. This process will convert the glTF files into game objects – more specifically, scenes in Godot terminology.

If you must fix something with your models or add detail, where can you do this? Since you are now in Godot, it's tempting to fix the models in Godot, but this is counterproductive. In this chapter, we'll show you how you can update your Blender file and reflect the changes in Godot.

In *Chapter 2, Building Materials and Shaders*, we learned how to work with materials in Blender. We'll revisit this topic in the context of Godot so that we can understand how materials work in both applications. We'll present the pros and cons of handling materials in either application so that you can decide which one works best for you. Whether you are working alone or in a team, there are a few decisions that can either save a lot of time or be frustrating down the line when you realize you've got to make a fundamental change. A decent material pipeline is one of these topics.

In *Chapter 5, Setting Up Animation and Rigging*, we stored two actions in our snake model. We'll import that model to see how Godot handles the animations stored in a glTF file. This chapter will only cover how to import animations; how to use imported animations will be covered later in this book when we build our point-and-click adventure game.

Thus, you'll be presented with some of the crucial building blocks and practices that will serve you in later chapters and your game projects.

In this chapter, we will cover the following topics:

- Making a scene!

- Going between Blender and Godot

- Deciding what to do with materials

- Importing animations

By the end of this chapter, you'll be able to take your glTF files and convert them into usable Godot assets, decide what to do with materials from a project pipeline perspective, and make sure you can access the animations that come with a model file.

Technical requirements

As mentioned in the *Preface* section, we assume you already know your way around Godot for basic things such as creating and composing scenes, adding scripts to **nodes**, using the **Inspector** panel to change the conditions of your game objects, and more.

However, if you are a novice in Godot Engine, then you may want to start with the official learning material at this address first: `https://docs.godotengine.org/en/3.4/getting_started/introduction/`.

Throughout this book, we'll be using Godot 3.4.4. There may always be something new or missing even between minor versions. Should you be using a different version when you are reading this book, you can either switch to the version this book is using or read the detailed changelog for different releases for the appropriate version listed at `https://godotengine.org/news`.

This is still a transitional chapter; as is the previous and the next chapter. We'll create a new Godot project in *Chapter 9*, *Designing the Level*, and work within that Godot project in later chapters to make a point-and-click adventure game. Until then, we can make do with temporary Godot projects. This means that in this chapter and the next, we won't be concerned at all with the structure of our files and folders. However, the sections in this chapter have been laid out in a way that we assume you are still working on the same Godot project.

As usual, this book's GitHub repository at `https://github.com/PacktPublishing/Game-Development-with-Blender-and-Godot` contains some files that are relevant to this chapter.

Making a scene!

In a typical 2D game built in Godot, using a **sprite** node is essential. You would then assign a texture to your sprite nodes in Godot's **Inspector** panel. The 3D version is essentially the same, but it involves using a **MeshInstance** node and then assigning a **mesh** to it. So, what textures are to sprite nodes is what meshes are to mesh instance nodes. Although creating a Godot scene that just has a sprite node and instancing this scene in a bigger scene is possible, it's overkill since you could easily attach the sprite node itself to the big scene.

This is where it makes sense to treat mesh instances differently and store them in their own scenes, unlike sprites, since 3D models have a lot more going on than getting assigned just one texture. Additionally, since a 3D model has a lot more moving parts, assigning individual meshes to mesh instances could be tiresome too, so let's do better. The goal of this section will be to create a scene out of a 3D model and to automate how to assign meshes to mesh instances.

> **Adobe Animate**
>
> Godot's scene concept incorporates a lot of notions you might be familiar with if you have worked with *Adobe Flash* in the past, or *Adobe Animate* these days, which uses **movie clips**, similar to what Godot does with its scenes. Creating nested movie clips and binding scripts is very handy, which is pretty much how a Godot project goes. Despite this similarity, there comes a moment in Godot when it makes sense to consider a 3D model its own scene, which is what this section will cover.

We suggest that you start a fresh Godot project for this section. Once you've done that, you need to find the `Sconce.glb` file in the `Start` folder for this chapter. You have two options at this point. First, you can copy and then paste this file, using your operating system's filesystem, into where your Godot project folder is. Alternatively, you can drag the sconce file to the **FileSystem** panel of Godot. When you make a glTF file as part of your project, you'll have something similar to the following:

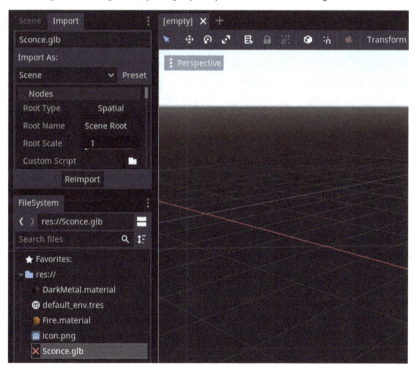

Figure 7.1 – The Sconce model is now part of your Godot project

Did you have a red cross as an icon for your Sconce.glb entry too? This doesn't happen all the time, but that icon indicates there is a configuration problem; luckily, the fix is easy. Restarting Godot fixes the issue most of the time. If that doesn't work, then we'll have to press a button to reimport the file for Godot to configure it for us. The preceding screenshot also shows the **Import** panel in focus. You can click the **Reimport** button at the bottom of that panel to make the file compatible with Godot.

Something else happened while we were discussing the icon issue. There are two material files in our project:

- DarkMetal.material
- Fire.material

These materials came within the glTF file that was exported from Blender since we opted to keep the materials. If you need a refresher on this, you can read the *Deciding what to export* section in *Chapter 6, Exporting Blender Assets*. By default, Godot will place the materials next to the model file. You may want to place your models and materials in separate folders for organizational reasons. We'll discuss something related to this in the *Deciding what to do with materials* section later in this chapter.

We're now ready to make a scene using the sconce model. This effort will create all the necessary bindings to display a Blender model in Godot. To achieve this, you must do the following:

1. Double-click the Sconce.glb entry in the **FileSystem** panel.
2. Click the **New Inherited** button on the pop-up screen.

The pop-up screen will display another button beside the one you've just clicked. There is also a piece of information about what each button does but it might be confusing, so let's explain it. In layman's terms, the **Open Anyway** button will let you see the contents of a glTF file, but this will be read-only. Since you may want to make alterations, such as attaching scripts, you'll often click the **New Inherited** button.

If you bring up the **Scene** panel, you'll see that your last effort has created two **MeshInstance** nodes under one **Spatial** node. When you click either the **Sconce** or **Flame** mesh instance node, you'll see their mesh bindings in the **Inspector** panel. We didn't have to create all this structure and bindings manually; creating a scene out of a glTF file did it all for us.

When you have finished examining what's changed so far, you can save your file as Sconce.tscn since it is still a temporary construct as far as Godot is concerned. The following screenshot shows our progress:

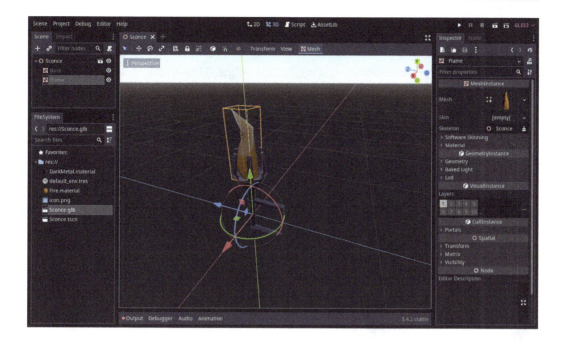

Figure 7.2 – You have created a scene with a click of a button

You can now utilize `Sconce.tscn` in other scenes by creating more instances of it. For example, in most 2D platformer games where there are enemies, you would have to create instances of scenes that stored the enemy character sprites. This is similar. Hence, every time you need a sconce, you can use the **Sconce** scene instead of the model file. We'll create many more instances of this scene when we work on our game later in this book.

Creating a scene out of a model file was easy, but how easy is it to change it? Either the sconce or the flame could use a bit of touch. We'll tackle how to update our models in our scenes next.

Going between Blender and Godot

In later versions of Godot, specifically starting with Godot 4.x, you'll be able to directly import Blender files into Godot and interact with them. Saving things in Blender will automatically update the situation in Godot. We are not there yet. At the time of writing, we must resolve to an already tried and tested method: re-exporting our assets. Let's see how we can accomplish this easily.

While you are developing your game, you'll likely want to make changes to your models. Perhaps you've been working with a prototype that your 3D artist friend or contractor provided a while ago. Now, they are ready to give you a more refined piece. So, let's simulate a similar scenario by making modifications to the sconce model we've been using. If you want to skip the Blender parts, you can find the finished changes in the `Sconce.blend` file in the `Finish` folder. If you want to exercise some Blender muscles, then we suggest that you make two changes in the `Sconce.blend` file in the `Start` folder. These changes are as follows:

- Move the tip of the flame so that it doesn't look too pointy. (Hint: go into **Edit Mode**.)
- Replace the flame material with something that is bright yellow. You can pick a name such as **HotFire** for it. (Hint: Remove the old material and add a new one.)

We're applying two important changes to our model. First, we are changing the geometry of our model, however minor it might be. Second, we are introducing a new material instead of changing the color of an existing material. All there is left to do is re-export our model and overwrite the existing `Sconce.glb` file in our Godot project. Chances are, if you were following along, the **Sconce** scene in Godot is still open, and despite overwriting `Sconce.glb` in the project, it looks like nothing has changed.

If you restart Godot, switch to a different scene tab, or do any other thing that would refresh the view, then you'll see your updates. Otherwise, you may still have the same old look. There is a general refreshing problem, it'd seem. Hopefully, little things like this will be fixed in future versions of Godot.

The following screenshot shows the updates you will see:

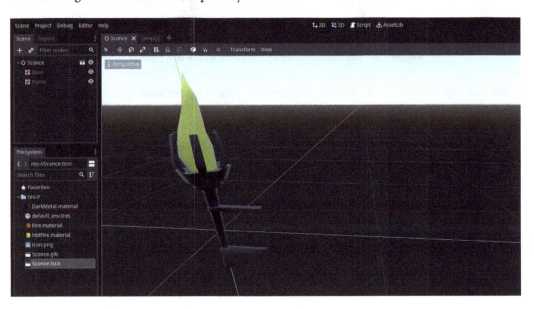

Figure 7.3 – The flame is hotter and straighter with our newest changes

While we successfully updated the **Sconce** scene, we also introduced a new material to the project. The import process was smart enough to know there was a new material coming in, but it was judicious enough to keep the old materials, just in case they might be needed and used sometime later in your project.

This could lead to having lots of unused files over time. That is not the worst of your problems, though. There is a much more insidious thing waiting for you when you import more and more models and eventually lose track of what's happening due to the sheer number of files as your project grows.

In the next section, we'll present a scenario where importing glTF files straightforwardly as we have done so far may cause some problems.

Deciding what to do with materials

An important decision awaits you. When you were exporting your Blender assets in *Chapter 6, Exporting Blender Assets*, we briefly discussed what the export options in the exporter's UI meant. However, we never really talked about the implications of keeping the materials or not. In this section, we'll present the pros and cons of handling materials in Blender versus Godot.

Let's assume you are now ready to import another model. For example, the `Vessel.glb` file in the `Start` folder is something you want to add to your game. If you take a look inside the associated `Vessel.blend` file, you'll notice that we are using a material labeled as **DarkMetal**. Ironically, perhaps accidentally, someone has decided to pick a light color, but the name, regardless of what the intentions are, is the same material name we used in the sconce model file.

So, what will happen when we import this file into Godot? To find out, follow these steps:

1. Add `Vessel.glb` to your Godot project.

2. Turn this vessel model into a scene. For familiarity's sake, save it as `Vessel.tscn`.

The following screenshot shows the new scene, as well as the status of the **FileSystem** panel:

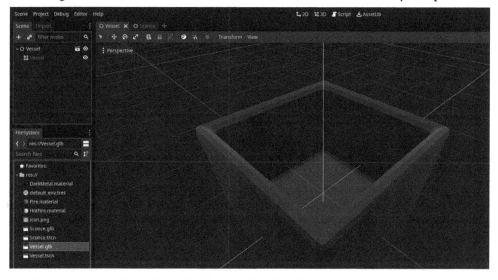

Figure 7.4 – Everything is kind of looking OK, but shouldn't this vessel have a lighter color?

Despite the mislabeling in Blender, we knew what we wanted for the vessel's color. It was supposed to be a lighter color but that's not what we are seeing in Godot. While importing the vessel model, since there was already a material with the same name in the project, Godot chose not to duplicate the resources. This is efficient, perhaps, but not accurate. This kind of thing could easily happen, especially if you are utilizing someone else's files. Luckily, only the new stuff doesn't look correct. In other words, the incoming resources are not overwriting and messing up the existing resources.

So, what do we do to make the vessel show the color we want? We can offer a few suggestions that are more organizational solutions. Thus, this is not a technical but a workflow or a pipeline type of solution as it is often labeled in the industry. Therefore, the solution lies in how you want to treat your files in your project, and whether you are working solo or in a group. These suggestions are to do the following:

- Label Blender materials by purpose
- Label Blender materials by color
- Import your models into separate folders
- Use a staging area in Godot

None of these suggestions is a magic pill. You have to try and decide if they're beneficial for you. Also, sometimes, projects of different sizes make some of these solutions easy or difficult to apply. The decision is yours after you learn what each one entails.

Labeling Blender materials by purpose

Naming materials in Blender by their shade, such as **DarkMetal**, can only go so far. How dark are we talking about? Sooner or later, we will find ourselves playing a game of adjectives: dark, darker, darkest, and likewise. It will get worse when we want to pick a lighter version of the dark tone we have already picked.

Typically, a sconce's base is wrought iron. Since it's a metal, it makes sense to use the word metal in its name, but it could easily get confusing. Instead, you could use the object's name for its material title. So, you'll have `Sconce.material` once you import it into Godot.

Labeling Blender materials by color

If you want to go with color-like labels, then you can make this obvious and in a unique way without leaving any room for Godot to interpret it in its own way. The **Hex** value in Blender for the **DarkMetal** material is **393646**. You could use that as a label. Hence, once imported, you'll have this material as `393646.material`.

Keep in mind that you'll often get busy and distracted while you are authoring your models and find yourself fine-tuning a lot of things in your models, whether it's geometry, materials, animations, and much more. So, if you have already chosen a hex color as a name, and later alter the color of the material, then you will have to remember to update the name.

Importing your models into separate folders

Some people organize their Godot projects so that they have separate folders for bigger concepts. This includes materials, models, scenes, and scripts. If you want to make sure your materials are unique to the model you are importing, a safer and easier way to do this is to create custom folders inside a specific folder. For example, if you have a `Models` folder at the root of your project, instead of dumping all the glTF files into this folder, you can create subfolders named after the model you are importing. In our case, this is the structure you'd see:

- `Models` > `Sconce` > `Sconce.glb`
- `Models` > `Vessel` > `Vessel.glb`

Then, all the relevant materials for each glTF file will be contained in their own folder. This might seem counter-productive at first since the same material file will be duplicated in different folders, especially if the material's names are color-coded. However, you'll at least know what you are importing is what you want in the first place.

This method might be advantageous in some scenarios. Maybe you are designing more than one sconce style for your game. In this new style, despite the wrought iron part having a different shape, it'll most likely use the same material. Then, you can easily rename the folder as `Sconces` to store multiple sconce files. This way, you're intentionally agreeing with the fact that Godot will not create duplicate materials but use the first imported model's material.

Last but not least, let's cover a caveat about this technique. If you are importing your files by dropping them over the **FileSystem** panel, you've got to be careful since that panel is context-sensitive. This means that you need to have the appropriate folder selected in the entry list. Otherwise, whichever entry is selected will be the recipient. To be sure of where you are sending your files, you can do all this by using your operating system's filesystem. When you switch to Godot, your files will be processed and, depending on the speed of your system, you might see a progress bar showing the progress of the import.

Using a staging area in Godot

Last in our list of possible solutions to making sure models and materials are imported properly is to use a staging area. This means, similar to using unique folders for models, you can designate a folder to monitor what's going on with a model. Perhaps this is a folder labeled as `Staging` inside the `Models` folder.

Using the search functionality in the **FileSystem** panel, you can even check if there are duplicate materials in other folders. This is a safe way to compare materials because you can observe their properties in the **Inspector** panel. If there are no obvious differences, and you deem it safe, you can just move the relevant glTF file to its final place while ignoring the duplicate material file in this staging area.

This requires a bit of work, but it might be a necessary practice in larger teams so that you can decide and even notify the artist if there are obvious labeling mistakes. For example, if there is a typo for the same material that multiple similar models should use, you won't end up with two separate materials.

Wrapping up

Out of all these options, and perhaps a few more you may find online, you must decide which one works best for you. It's a common thing that you'll start one way and switch to an alternative method as your project's needs change. Although your choice might have technical ramifications, it's rather a business decision; so, weigh up the pros and cons while making it.

There is one more thing you can do regarding your materials and models, but since this chapter is about importing, we're intentionally leaving it for later. It's when you decide to create your materials in Godot and bind them to the meshes of a model manually because, sometimes, you find models with just their meshes but with no material information. We'll show you how to create materials in Godot in *Chapter 9*, *Designing the Level*, in the *Constructing the missing materials* section.

Now that we seem to be done with material things, in the next section, we'll learn how to import the animations we created for our snake in *Chapter 5*, *Setting Up Animation and Rigging*.

Importing animations

The last thing we'll cover about importing Blender assets is animation. Thus far, we have taken care of importing the mesh and materials of a model. We even discussed workflow problems concerning the default import workflow for materials. Hopefully, there won't be a hidden surprise in importing animations, but how do we do it? You'll find out in this section.

You can start by moving the Snake.glb file in the Start folder of this chapter to your project. Then, as shown in the *Making a scene!* section, you can create and save a scene out of this model. The snake model will bring a lot of its materials, and your **FileSystem** panel will look a bit crowded, but this is what we have so far:

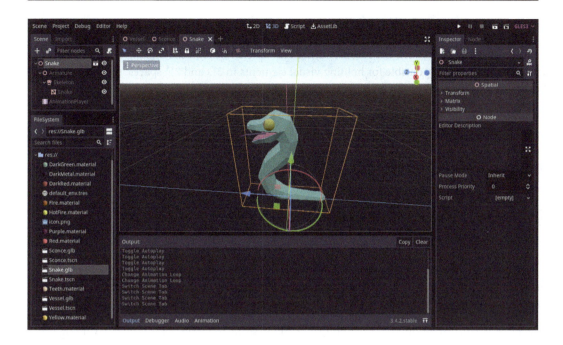

Figure 7.5 – The snake has followed you to Godot Engine

This is a good time to introduce some of the 3D nodes Godot uses. We'll utilize the snake scene for this effort because it has a good sample of different nodes you'll most likely use in your projects.

The root node is of the **spatial** type. Godot's chosen color for all 3D nodes is red. If you've been using Godot's **Node2D** nodes, which have a blue circle icon, **Spatial** nodes are the equivalent of that, only in 3D, and they are red. Whereas a **Node2D** node will have coordinates only in the *XY* plane, a **Spatial** node will have them in the *XYZ* plane. You usually employ this kind of node as a root container for other nodes. For example, the **Armature** and **AnimationPlayer** nodes are the direct children of the root **Spatial** node, which is labeled as **Snake**.

> **Node types versus labels**
>
> In the snake scene, the **MeshInstance** node has been renamed **Snake**, which is useful so that you can differentiate if you have a lot of mesh instances. There is no built-in *Snake* node type in Godot, but it's okay to just say **Snake** node, even though it's of the **MeshInstance** type. The **Inspector** panel will figure out the type and only list the relevant properties. Therefore, throughout the rest of this book, we'll refer to the scene's nodes either with their custom labeled names or node types.

We'll soon analyze the role of **AnimationPlayer**, but let's finish looking at the **MeshInstance** and **Skeleton** child nodes first.

MeshInstance and Skeleton

We made an analogy between the **MeshInstance** and **Sprite** nodes in the *Making a scene!* section, stating that they are responsible for holding visual elements in 3D and 2D spaces, respectively. So, that leaves us with the **Skeleton** node.

In *Chapter 5*, *Setting Up Animation and Rigging*, we used bones and attached them so that we could animate the snake. When the `Snake.glb` file was imported, the bones were imported as a single unit. In other words, Godot grouped all your bones into a node of the **Skeleton** type. However, you can still access each bone if you wish:

1. Select the **Skeleton** node.

2. Expand the **Bones** section in the **Inspector** panel.

3. Expand some of the entries, especially **9** and **10**.

Do you recognize the names? These are the names you picked for the bones in Blender. Look at how much preparation we require to construct a skeleton. The rigging process to create all this, however complicated it may have looked initially, is still far too easy to do in Blender compared to Godot.

Now, let's turn our attention to the last node type in the scene to further appreciate why doing the animation in Blender was also a superior and preferred move. Enter **AnimationPlayer**.

AnimationPlayer

The last node in the **Snake.tscn** scene is **AnimationPlayer**. The color of this node is neither blue nor red. This means you can use it in both 2D and 3D contexts. You may already be familiar with this node if you have been building 2D games. If that's the case, then you know that you need to place keyframes in the player's timeline to mark the changing points, just like we did in Blender. Regardless of whether you have experience with **AnimationPlayer** or if this is the first time you are tackling it, you'll notice that creating so many keyframes, as shown in the following screenshot, is a lot of work:

Figure 7.6: You worked smart, not hard, to create all those keyframes in AnimationPlayer

Each orange diamond in the preceding screenshot is a keyframe and marks an important turning point in the animation's life cycle. This is the timeline for the **Attack** action we created in Blender. You can see it in a dropdown in the top section. This is the main reason why we opted for Blender to create all this for us – we were only concerned with the major events, not with what exactly happened in between major events. Godot and Blender worked together to fill in the details. Also, updating your animation in Blender is still a much better idea than fiddling with those diamonds.

As you can see, animations and actions are automatically imported, recognized, and organized in **AnimationPlayer** for us. Despite how easy this was, there is currently a bug in Godot regarding the animation imports. So, we need to do something about it that may not be necessary in the future. We'll discuss what the problem is and present a solution here. However, to follow the discussion and updates on the problem, you can go to `https://github.com/godotengine/godot/issues/34394`.

On the right-hand side of the **Animation** panel, there is an icon that looks like a recycling symbol. At the time of writing, that loop button, which is supposed to play an action indefinitely, only works while you are editing a scene. So, even though you can toggle the loop button on, the action will play only once when you launch the game. Hopefully, soon, newer Godot versions will fix this looping issue. Still, it makes sense to present a workaround for the time being.

Separating actions

Luckily, there is a solution to the problem we've just presented. We'll instruct Godot to separate the actions into separate files, similar to the way materials for a model are kept in the filesystem.

The default behavior for keeping animations for a model is to store them inside its file. In this case, the `Snake.glb` entity is holding all its animations. To extract these animations, follow these steps:

1. Select `Snake.glb` in the **FileSystem** panel.
2. Switch on the **Import** panel and scroll down to the **Animation** section.
3. Choose **Files (.anim)** in the **Storage** drop-down options.
4. Click the **Reimport** button.

The following screenshot shows the steps we have taken so far:

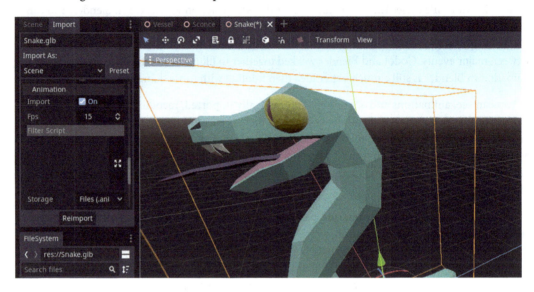

Figure 7.7 – The import settings for the Snake.glb file

This will extract the actions into the filesystem. In the end, you'll have two more files in your project:

- `Attack.anim`
- `Idle.anim`

These are the actions you defined in Blender a while ago. Also, similarly named actions are listed in Godot's **AnimationPlayer** dropdown. For example, *Figure 7.6* shows the **Attack** action selected. There is one more step left for us to fix the looping issue – it's to reintroduce these actions we've just separated back to **AnimationPlayer**, even though it's already listing them. To achieve this, follow these steps:

1. Switch the **Scene** panel on.
2. Select the **AnimationPlayer** node in the scene structure.
3. Click the **Animation** button in the **Animation** panel (the button to the left of the **Action** dropdown).
4. Choose **Load** and select `Attack.anim` from the **Open a File** pop-up menu.
5. Repeat *Step 4* to load `Idle.anim`.

This will replace the existing actions with the actions coming from your filesystem. The following screenshot shows where you can find all these names since there have been a lot of similar words. Here, the **Animation** button has already been pressed and is displaying the available commands:

Figure 7.8 – The Animation panel's menu for loading, saving, and doing many other things

In the future, hopefully, you won't have to separate and reimport your actions with newer versions of Godot. For the time being, this will work, but we won't see the effects of this until we get to the later chapters of this book, where we will trigger these actions.

Summary

Since we took care of exporting Blender assets in the previous chapter, it was time to learn how to import these into Godot. This is what we covered in this chapter.

First, we learned that once a glTF file is part of a Godot project, Godot automatically takes care of things such as separating materials. That being said, since we'd most likely keep creating more instances of 3D assets, we looked into creating dedicated scenes out of glTF files. Moreover, we learned how to make modifications to our models in Blender and get the scenes using these models updated back in Godot.

Then, we covered materials, which is an enmeshed topic within the model workflow, and discussed different ways of labeling the materials, and even keeping the models in separate folders to prevent any material file from overlapping. You decided what works best for you since this kind of thing might be team-size or project specific.

Finally, we tackled how easily animations can be imported. Creating a scene out of a model took care of all the scaffolding. Even though we'll learn how to trigger animations in later chapters, especially for looped animations, we presented a problem that may occur. A workaround was presented, and we hope you won't need this in the future.

This was your first chapter on Godot and you're now officially using Godot Engine. Importing 3D assets into Godot is an essential operation, and we hope you have a seamless back-and-forth between Blender and Godot for your games.

In the next chapter, we'll still work on a standalone topic, *Adding Sound Assets*, to keep things simple. By the end of the next chapter, we'll have covered the basics of setting up a project structure, which means we can focus on building the game after that.

Further reading

You've already interacted with the **Import** panel of Godot. That area has a lot of settings that would require us to write a chapter to investigate all possible combinations. The default settings work most of the time but there is a **Preset** button in the top-right corner that lists the most used combinations.

Since the needs of a project, and thus the import requirements of a model, won't be clear ahead of time, we leave the task of discovering what those options entail to you. That being said, here is the official resource that can guide you if you want to get more information: `https://docs.godotengine.org/en/3.4/tutorials/assets_pipeline/importing_scenes.html`.

Similarly, you may want to import images instead of 3D assets. This is necessary when you are building UI elements for a game. We're mostly covering the 3D workflow throughout this book, so we won't emphasize the import settings for 2D assets. Nevertheless, if you want to be informed before we tackle the UI topics, here is the official URL: `https://docs.godotengine.org/en/3.4/tutorials/assets_pipeline/importing_images.html`.

8

Adding Sound Assets

Sound is often the most neglected part of game projects. While creating visual assets may seem hard to do, a lot of us still tackle it because we get quick and reliable feedback, however, most people don't even know where to start when it comes to producing sound assets. Luckily, there are royalty-free assets out there that you can use.

This chapter will not cover how to make sound assets but how to import them into your game. We will focus on some of the technical aspects of sound management in Godot. This involves learning about the different sound formats the engine supports. Picking the appropriate sound format is no different than ironing out a topology for a 3D model for animation. Choose wisely and, even better, know the benefits and limitations of each format.

Next, you will learn when and how some sound assets should be looped. We'll investigate the import options for different sound types and mention format-specific differences. We'll also discuss scenarios where it makes sense to have your sound assets looped.

Lastly, we'll get to know different types of Godot nodes that are responsible for playing sound assets in your scenes. This way, you can pick the appropriate audio player node for your project. To finish off, we'll play some sample sound assets to show the differences between these different nodes.

Needless to say, to make the best of this chapter, you may want to be in a quiet place where you can practice some of the topics, especially in the later sections of this chapter.

In this chapter, we will cover the following topics:

- Learning about different sound formats
- Deciding on looping or not
- Playing audio in Godot

By the end of this chapter, you'll know how to import sound assets, choose which file type is correct, configure their settings, and play them in your project automatically or when it's needed.

Technical requirements

Unlike the other chapters, instead of a `Finish` folder with individual assets, we'll give you the finished Godot project with all the scenes and scripts set up. Nevertheless, we would like you to practice but focus solely on the topics presented in this chapter. Thus, we suggest you start with a clean slate, import the sound files from the `Start` folder, and follow along. Following tradition, the necessary resources can be found in this book's GitHub repository: `https://github.com/PacktPublishing/Game-Development-with-Blender-and-Godot`.

Learning about different sound formats

Sound files come in different formats, just like graphics files can come in different formats including JPG, GIF, PNG, and others. The industry, and sometimes the consumers, define the fate of these formats. Let's place the consumers in the right context here. Occasionally, the specifications laid out by the creator of a file format are not welcome by the people who are using this very format to produce the work. Then, the work is created but not accepted by the platforms that would disperse such content due to technical reasons. It's almost like a tug of war where the inconvenience or the cost of maintaining a file type outweighs the benefits and the ease of use. At these times, we tend to hear about newer formats, hence there being a multitude of file formats out there.

Most of the time, this kind of technical layer is not visible to an end user, especially if they are only perusing the content, such as listening to music on Spotify or YouTube. However, since we are building a game, even though we are not too concerned about the production of such assets, we should still be knowledgeable on this topic since we'd like to pick the most appropriate file format for a certain scenario.

> **Distinguishing what sound means**
>
> This is a note on what we mean by sound. We'll be using the word sound or audio, in this chapter and the rest of this book, to cover all possible scenarios, such as the feedback you get when you interact with UI elements, when a player character is notified by an in-game event, or ambient music.

The version of Godot, 3.4.4, that this book is covering currently supports three different audio file formats. Each has different advantages and limitations. Although converting these files into each other is possible, after we present their formal definitions, perhaps you'll decide not to.

Introducing WAV

Pronounced *wave*, WAV files have been around since the early 90s. It's the short form of Wavefront Audio File Format, a file specification created by IBM and Microsoft. This is a popular format among music and audio professionals, despite being uncompressed since it retains the quality of a sound recording. Thanks to the improvements in file storage capacity and internet speed, the high file size doesn't seem to be a big issue anymore.

On the limitation side, as far as technical aspects go, a WAV file can't exceed 4 GB. However, this should not be a concern because that number is equivalent to almost 7 hours of audio. It is extremely unlikely there will be one audio file in any video game of that size.

So, why should you choose this format? Since it's an uncompressed file type, the CPU that is also responsible for processing a sound file will have an easier time playing it. A likely scenario for using this file type is for sound effects. Usually, these effects are short-lived, such as the creaking of a door, the swing of a sword, and so on. The file size won't matter that much because the duration will be short.

Conversely, this is not the best format for background music. Sure, there won't be any need to decompress the file to be able to play it, but the file size will be significantly larger.

In summary, if you want a quick reaction and would rather have a sound file play as quickly as possible, such as effects, then this is the right format for you. After all, you wouldn't want the CPU to be dealing with the decompression of an effect file while your game characters are busy with the next chain of events.

If you are willing to sacrifice a few hundred milliseconds to wait for a decompression, such as when not having the background music play instantly is a big deal, then you can opt for compressed file types. These come in two different flavors.

Introducing OGG

We should start by clarifying this format since the name could be confusing if you come across some resources on the internet. Technically, OGG is a container file format that can hold file types such as audio, video, text, and metadata. Its developer and maintainer, *Xiph*, is also responsible for another audio file format known as **Free Lossless Audio Codec (FLAC)**. So, according to OGG specifications, a FLAC could be part of an OGG file. Historically speaking, most OGG files out there have contained a different audio file format known as Vorbis. So, you may find some websites with Vorbis content that are essentially complying with the OGG format's specifications.

Here is an example to simplify all these names and how they relate to each other. Consider OGG as a ZIP file that knows what to do with its content. An OGG file carrying a video and a subtitle file will trigger the necessary settings in a video player so that the player knows where to find the subtitles since they will be embedded in one file. Similarly, another OGG file with an audio and metadata file will command an audio player to display the album and track, record, and play the audio.

Since the format is not just one thing, but rather a set of files, it is often confusing to associate a specific need with one file extension. For example, the .ogg extension was used before 2007 as a multimedia holder as that was its original intention. Since then, Xiph suggests we use the .ogg extension for Vorbis audio files. Additionally, the company has created a new set of file extensions to simplify things:

- .oga for audio-only files
- .ogv for video
- .ogx for multiplexed cases

Despite the naming conundrum, what you need to know is that the OGG audio format is compressed, so it's a lossy file format. Lossy in our context means that we could attain almost the same sound quality by requiring less hard disk space. So, this is a good thing because this file format is a perfect fit for playing background music. Keep in mind that since the CPU has to decompress this file type, this is not the preferred format for playing quick sound effects.

Speaking of a lossy file format, our next candidate is another lossy file format that gained some notoriety in the early 2000s.

Introducing MP3

When internet speed and disk storage were at a premium in the late 90s, MP3 filled an important gap in transferring audio content just when a big audience needed it at the turn of the millennium. Consumers flocked to websites to download copies of the tracks from their favorite bands. Sadly, so many of these websites did not bother to have a legal license to distribute such content, so this led to copyright infringements and, in the case of Napster, a lawsuit.

From a technical standpoint, MP3 files are somewhere in between WAV and OGG, compression-wise. So, you'll get smaller file sizes in OGG for the same quality of sound. That being said, decompressing an MP3 file is faster than decompressing an OGG. Hence, this makes the MP3 format still useful, especially where CPUs are challenged to the maximum, such as in mobile devices.

Despite disk space getting cheaper and cheaper, from a business point of view, it still makes sense to prioritize WAV over MP3. For example, some websites that offer royalty-free sound files provide the MP3 version but put the WAV version of a sound behind a paywall. Since an MP3 file has already lost some of the original data due to its compression algorithm, editing with this file over and over will yield more lossy results. So, having access to the original WAV file is always better if you want to make modifications to it. However, if you don't need to, then you might be fine with an MP3 version.

Wrapping up

In summary, WAV files are better for short sound effects whereas longer sound effects, especially theme music, would be handled better with MP3 files. At the time of writing, most sound libraries still don't offer OGG commonly, despite being a good candidate. Nevertheless, if you have access to a lot of WAV files and you want to be efficient in file size, then you can convert them into OGG using online converters. Two examples are as follows:

- `https://audio.online-convert.com/convert-to-ogg`
- `https://online-audio-converter.com/`

In the case of music files, which are normally a few minutes long, if your original is in WAV format, then uploading and processing these files online may take a long time since the file sizes will easily be over 50 MB. Also, some of these online converters have file size limitations. To get around these limitations, here is a link to a website that compares some offline converters that you can employ in your efforts: `https://www.lifewire.com/free-audio-converter-software-programs-2622863`.

Regardless of what file type you choose and whether it's for a sound effect or music, there comes a point in your game development journey when you will have to decide if your sound asset should loop or not. In the next section, we'll discuss the reasons why having the loop feature on or off is useful.

Deciding on looping or not

A loop, in literal terms, is a continuous motion or structure in which if you pick a random spot, you could come back to it by traveling all the way through. In aural terms, this is similar, but we don't start anywhere; we usually start playing a sound file, but the player restarts the track once it reaches the end.

This definition is classic, and not that insightful, so let's do a better job by discussing it in various contexts inside Godot or any game projects. So, you can make informed decisions in your projects since it's situation-specific. We'll do this by presenting different use cases:

- **Background music**: This is the most typical case where a music piece plays in the background while the game is running. The composer creates this kind of piece with the intention that once played back to back, there will be no abrupt end. The sound at the end of the file will seamlessly match the beginning. Sure, if you pay attention to the ups and downs in the rhythm, you will know where you are in the file, but so long as the loop setting is on, everything will sound smooth and blend in so that you can focus on your game experience.

- **Machine gun**: Imagine that either the player or an enemy character is interacting with a machine gun in your game. Although short bursts are possible, due to the nature of machine guns, the gun might be fired continuously. So, instead of detecting if the sound file has reached the end and instructing the player to restart the file, you may want to play the file once if the said file's loop feature is on. This way, the machine gun effect will play until a stop command is given.

- **Doors**: This one is a bit of an edge case. Let's assume we have visuals and other sound effects in our game that indicate that we're in an outdoor scene on a windy day. Perhaps the door is in poor condition with rusted hinges, and one of the hinges is even leaning out a bit. The artist may have decided to have this door animated to match the wind's effect on the door so that it oscillates between a closed and an open state. Here, it would make sense to have a looped sound file that contains most likely squeaks and creaks that are synchronized with the door's animation.

  However, if a door will be responding to a player character's action such as it being opened or closed, then it doesn't make sense to have the sound file in a loop. This is going to be a one-off event.

- **User interface**: The sound you hear when you interact with a user interface falls under this category. These are usually not looped since they are event-based, similar to the one-off-door action from the previous use case. However, let's present a case that may seem like looping is a good idea. Nevertheless, we'll rule it out for a good reason.

Imagine that there is a UI component that's helping the player set an amount. The interface has two buttons that will increment and decrement the amount the player is seeing. Placing a UI sound effect on either button is fine, and the sound will play only once, so long as the player keeps clicking. What if we would like to give the player a chance to press and hold the button down? After all, clicking a button ad nauseam to get to really high or low numbers may get tedious quickly. So, how should we treat the looping condition in this case?

Human perception is sensitive during events like this. Players are usually busy during gameplay, so they won't perceive the delay while the CPU is busily decompressing a music file. However, we are usually very perceptive in detecting the discrepancies at the end of a holding event for a UI button. So, instead of treating repetitive UI events such as a machine gun, even though they might feel similar, designers opt to trigger the sound effect individually instead of looping it.

In this section, we presented different use cases where the use of looping, or lack thereof, is common. However, what you haven't seen is how to turn the loop functionality on and off. We'll show this by revisiting our old friend, the **Import** panel.

Turning the looping on and off

So far, we have discussed what looping is and under which scenarios it may make sense to have it on or off, but we haven't seen how we can flip its status. In this section, we'll put sound files of each type in our project and study their settings in the **Import** panel.

We are going to use the `Loop_Someday_03.wav` file from the *Freesound* website, which was created by a user called *LittleRobotSoundFactory*. The sound was originally in WAV format, but we have converted it into OGG and MP3 versions as well. You can find all the versions in the `Start` folder and compare their file sizes.

Once you've added the files to your project, let's learn how Godot recognizes these files. So, switch on the **Import** panel, and select either the OGG or MP3 version. Then, select the WAV version. The interface differences are shown in the following screenshot:

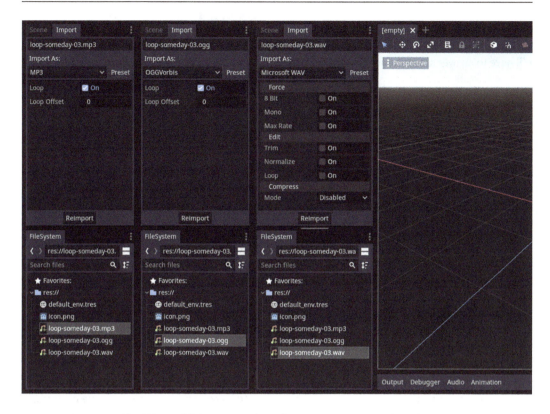

Figure 8.1 – The MP3 and OGG versions have fewer import settings than the WAV version

As you can see, by default, the MP3 and OGG versions come with the loop setting on. Also, these versions don't seem to have that many settings. On the other hand, the WAV version's loop is off by default. Why is that?

If you remember what we introduced for different sound formats earlier in the *Learning about different sound formats* section, Godot took the liberty of looping the compressed versions since these will most likely be used for background music. On the contrary, if our example file was for a sound effect, we'd most likely use a WAV file with no loop, since it'd be a quick one-off thing with minimal CPU requirements.

> **Other WAV settings**
>
> Since we are currently working with the **Import** interface, let's also point out that you can reduce the file size of your WAV files by turning on some of the options in the **Force** section. *Figure 8.1* shows this and some other settings, such as trimming and normalizing your files. The former of these will trim the silent part at the beginning and the end of files, which is sometimes automatically added when exporting WAV files. This is especially important if you want your sound effects to start right away without a delay.

So, turning the loop feature for any given sound file on and off is as easy as a click and you know how to do it. Perhaps it's more important to decide whether a file should be looped or not. This is something you'll have to answer along the way.

Regardless, you still need a Godot node to play your sounds at some point. In the next section, we'll get to know the different audio players Godot uses, and attach our sound files to the appropriate player.

Playing audio in Godot

Since Godot uses nodes for almost everything, it is no different for playing sounds. To play an audio file, there are nodes you can attach to your scene, and you can configure them according to whether it's for a 2D or 3D game. We'll focus on different audio players Godot uses in this section.

No matter what audio file type you choose, you will be able to play it with the nodes we'll present in this section. The experience you'll feel will be different, of course, based on the node type, but this is something you have to decide, depending on the type of game you are making. So, let's look at the audio streamer nodes Godot uses so that you can pick the appropriate one. Your three choices are as follows:

- **AudioStreamPlayer**: This node's official definition is somewhat dry; it plays audio non-positionally. What this means is that you are not concerned with which direction the audio is coming from. For an FPS game, it's essential to know in which direction the enemy is firing at you. This involves positional data. You don't have any kind of positional information in this audio node. However, this is the right candidate for playing background music. Find more about it at https://docs.godotengine.org/en/3.4/classes/class_audiostreamplayer.html.

- **AudioStreamPlayer2D**: You guessed it – this node includes position information. So, the farther away the camera is from this node, the quieter the sound will be. This node is useful for 2D platformer games, for example. So, as soon as a game object enters the view, the stream will be picked up by the camera. Also, objects that are on the right-hand side of the camera will prioritize the right speakers and vice versa. More details are available at https://docs.godotengine.org/en/3.4/classes/class_audiostreamplayer2d.html.

- **AudioStreamPlayer3D**: Last but not least is the 3D version of an audio streamer. This conveys 3D positional information to a listener. Therefore, this is the kind of audio streamer node you'll be using in 3D setups. Naturally, this type of streamer employs more advanced features, such as attenuation, which controls how the sound will dampen over a distance, and Doppler effects. Thus, it might be a good idea to examine its properties by visiting https://docs.godotengine.org/en/3.4/classes/class_audiostreamplayer3d.html.

We could go over every property for each type of stream player, but we leave that task to you since picking the right streamer and configuring its settings is a form of art. We'll use the proper streamer when we build our game later in this book and focus on the important settings in that context. In the meantime, you can read what each one is capable of by going to the aforementioned URLs from the official documentation.

That being said, we won't leave this chapter just yet. Let's play a few sounds to simulate some of the examples we've enumerated so far.

Playing background music

Let's practice some of the things we've covered in this chapter. We'll start by playing a sound that's a good candidate for background music. We'll use the MP3 version of the `loop-someday-03` file we imported in the *Deciding on looping or not* section. To play this sound as background music, follow these steps:

1. Create a new scene and save it as `Background-Music.tscn`.

2. Add an **AudioStreamPlayer** node to your scene and turn on its **Autoplay** property in the **Inspector** panel.

3. Drag and drop `loop-someday-03.mp3` from the **FileSystem** panel into the **Stream** property in the **Inspector** panel.

4. Press *F6*.

This will launch your current scene and automatically play the MP3 file. Since the file's loop setting is set to true, the 9-second-long music will play endlessly. You can now add this scene to other scenes where you want to have background music.

Playing a sound effect on demand

For this effort, we'll return to the machine gun example from the *Deciding on looping or not* section. The sound for the machine gun is also set to loop, but we wouldn't want this to *autoplay* when a scene is launched. It's most likely that your player character will enter or approach an area where enemy forces are pummeling you with machine gun fire. Let's write some code to simulate this sort of triggering behavior:

1. Create a new scene and save it as `Machine-Gun.tscn`.

2. Add an **AudioStreamPlayer** node to your scene and attach a script to it with the following lines of code in it:

```
extends AudioStreamPlayer


func _unhandled_key_input(event: InputEventKey) -> void:
    if event.is_pressed() and event.scancode ==
      KEY_SPACE:
        play()
    else:
        stop()
```

3. Drag and drop `machine-gun.ogg` from the **FileSystem** panel into the **Stream** property in the **Inspector** panel.

4. Press *F6*.

Since we want the stream to play on demand, we are wiring it to a condition to be true – that is, pressing the spacebar. Go ahead and press it once or twice; even hold it down for a brief period. You'll hear the machine gun sound going on or off, thanks to the play and stop commands of the **AudioStreamPlayer** node.

The script we've implemented looks good enough, but it's also a bit problematic. Maybe you've already noticed it. Try to hold down the spacebar for long enough, such as 3 or 4 seconds, and you'll hear a jamming sound. This is because the script is firing too many play commands. So, after a while, the CPU will be instructed to play the same asset too many times. We can do better by replacing the script's content with the following:

```
extends AudioStreamPlayer


func _unhandled_key_input(event: InputEventKey) -> void:
    if event.is_pressed() and event.scancode == KEY_SPACE:
        stream_paused = false
    else:
        stream_paused = true
```

Here, we have replaced the lines that had the play and stop commands with a different kind of command. The new version will control whether the stream should be paused or not. For this script to work, we need to turn on two things in the **Inspector** panel:

- **Autoplay**

- **Stream Paused**

This new setup will play the stream automatically, similar to what happened in the *Playing background music* section, but then pause it right away. This seems counter-intuitive at first, but let's analyze what the new script is doing. When the spacebar is pressed, we resume the stream, and since the stream was already playing, thanks to **Autoplay** being on, you get to hear the ta-ta-ta-ta sound! Also, when you release the spacebar, hence the `else` case, the stream will be paused again. So, the new script will not send consecutive play and stop commands, and thus will not clog the CPU.

We'll conclude by discussing two more flavors of the machine gun firing in light of what we have presented throughout this chapter.

Increasing gameplay experience

Did you notice that we used the same type of audio stream node for both the background music and machine gun? In a way, we treated the machine gun as if it was background music. In other words, we were not too concerned about where the sound was coming from.

To deliver a more enjoyable gameplay experience, you could use the **AudioStreamPlayer2D** and **AudioStreamPlayer3D** nodes in 2D and 3D games, respectively. By tweaking the attenuation values of these nodes, which define how sound travels over distances, your players can hear the sound of the machine gun louder and louder as their characters get closer to the source. This would elevate the sense of danger, and it's a cheap and nice way to deliver immersion.

Summary

We started this chapter by presenting different types of files that Godot uses for playing sound. Knowing the differences among these formats, when you work with composers, you can emphasize in which format you want your sound files to be delivered. Chances are they might ask you about this, and they might even deliver in all three possible formats.

Next, we discussed a few cases where looping a sound file might be a good idea. To facilitate this, we investigated the options presented in the **Import** panel. However, the decision to loop or not is still something you'll have to decide.

Finally, to put our theoretical knowledge to use, we created two scenes that could play the sample files. In the first case, we attached a sound file to an audio streamer and let it play automatically. For the second case, we wrote a very simple script that let you start and stop the sound to mimic an enemy character's behavior, hence the sound effect it may make.

So far, we have been discovering some of the ingredients that are necessary for building games, such as importing assets – whether it's models from the previous chapter or sound assets in this one. In the next chapter, we'll dive right into building our point-and-click adventure game by designing our level.

Further reading

If you are into creating music and sound effects, here is a short list of software you can start with:

- LMMS: `https://lmms.io`
- Waveform Free: `https://www.tracktion.com/products/waveform-free`
- Cakewalk: `https://www.bandlab.com/products/cakewalk`

The aforementioned links will only cover the using a tool side of music production, so you may also need to learn the artistic side of it, for which there are courses on multiple online training platforms, such as Udemy.

By the way, if you see a sound file out there and it looks like it is free to download, it doesn't mean you have the license to utilize the piece in your work. You may want to read the fine print if you don't want to get a surprise call from a lawyer someday. Nevertheless, the following are a few websites that offer paid and free sound content:

- `https://gamesounds.xyz`
- `https://freesound.org`
- `https://www.zapsplat.com`
- `https://opengameart.org`

Part 3: Clara's Fortune – An Adventure Game

In this final part of the book, you'll be creating a point-and-click adventure game. Since it would be too time-consuming to prepare all the game assets, you'll be provided with the necessary files.

In this part, we cover the following chapters:

9
Designing the Level

From this chapter on to the end of this book, you'll be actively working on creating a point-and-click adventure game. We'll show you the necessary steps to create a game in which you'll place and command a character whose name is Clara. Players will be controlling her actions inside a cave that will be initially dark, but you'll be able to give controls to the player to change the conditions of the lights. Once you figure out how to move her around in the world, you'll also place trigger points in this cave so that the world reacts to Clara's actions to make things interesting but also challenging. This part of this book will cover enough basic building blocks for you to start practicing building small-scale adventure games.

Through all these efforts, you'll learn how to utilize different parts of Godot Engine, especially the ones that are pertinent to 3D workflow. Whenever it's necessary, we'll remind you of the previous chapters, where you can revisit some of the basic principles. This is because this part of this book will heavily rely on practical applications of what we have presented so far.

With that said, as every game has a narrative; this is ours:

"It was no more than a fortnight ago when Clara's uncle had sent for her. Clara was sailing her boat to the coordinates her uncle gave her when she noticed a glimmer in the distance. After she carefully approached the spot where she noticed the flash, she saw that this was the entrance to a cave under the cliffs of a rock formation jutting out of the sea. She cautiously maneuvered the sails on her boat and entered the cave without a hitch. Luckily, there was enough sunlight for her to see a pier and she anchored the boat. She's excited to visit her uncle."

Although there is a lot to do, from adjusting the lights in a cave environment to triggering sound and animations, we should start building the world first. That's what this chapter is about.

We'll start by composing a scene by placing models from the project folder. This kind of scene structure, where the players experience a particular part of the game world, is often called a **level** and often signifies different levels of difficulty or a distinctive environment.

While we are arranging assets to build a level, we'll look into creating and fixing materials in Godot since, sometimes, some things are not perfectly transferred between applications. *Chapter 6, Exporting Blender Assets*, and *Chapter 7, Importing Blender Assets into Godot*, covered the intricacies of how exchanging information between Godot and Blender works if you need a refresher.

Although manually laying things out to create a level is alright, we could always benefit from using tools that will make this kind of job easier on us. Godot's **GridMap** is the right tool for placing objects on a grid structure. For **GridMap** to work, it needs another Godot mechanism called a **MeshLibrary**. We'll show you how to construct one and use it as an alternative way of building levels.

In this chapter, we will cover the following topics:

- Creating the cave
- Constructing the missing materials
- Laying models on a grid
- Taking advantage of MeshLibrary

In the end, we'll craft a level by arranging scenes/models, completing missing materials, and taking advantage of **GridMap** and **MeshLibrary** for a faster workflow. By doing this, you'll have the right tools under your belt to design levels.

Technical requirements

Starting with this chapter, and continuing in the remaining chapters, you'll be creating a point-and-click adventure game. Since it'd be too time-consuming for you to prepare all the game assets, we are providing them. We have already exported the glTF files from Blender. Should you need to access the originals for any modifications, or when a specific file is mentioned, these files can be found in the `Blender Models.zip` file in this book's GitHub repository.

Unlike the previous chapters, which usually had `Start` and `Finish` folders with simple assets, we'll switch things up a bit. This chapter will have the usual folders too, but they will contain the content of a Godot project. The Godot project in the `Start` folder will contain the barebone assets for you to start building the level for the game. By the end of this chapter, your game will have reached a stage where you can use the content from the `Finish` folder to compare what you have created.

Additionally, starting with the next chapter, you'll only have the `Finish` folder since you can use the finished stage in each chapter as the starting condition for the following chapter, and so on.

We suggest that you head to this book's GitHub repository at `https://github.com/PacktPublishing/Game-Development-with-Blender-and-Godot` to check out the content we have prepared for you and help Clara out in her adventures.

Creating the cave

For the first level in Clara's adventures, we thought of a small place so that you don't get overwhelmed with building a large layout. *Figure 9.1* should help you visualize what we are building. This is a Blender render we'll try to recreate in Godot:

Figure 9.1 – We'll be building this small level for Clara to discover

Our world will consist of a dock inside a cave that has access to the sea. When Clara anchors her boat, she sees inside the cave. There isn't much light to begin with, but as little as she can see, the dock leads to a pier with laid stone. She can also see that there are a bunch of boxes, barrels, and pots distributed here and there. Though the sconces on the walls will start unlit when the game runs, as shown in *Figure 9.1*, you can see that all the sconces on the walls are lit. This is because we want to show you a later stage in the game so that you can see what we are aiming for. Otherwise, it would have been a dark figure.

In *Chapter 10*, *Making Things Look Better with Lights and Shadows*, we'll investigate how we can create a more dramatic-looking level by utilizing appropriate light types and enabling shadows. We covered some of this in the context of Blender in *Chapter 4*, *Adjusting Cameras and Lights*, but we'll do it in the context of Godot as well.

> **Level design versus game design versus visual design**
>
> If you are new to game development, then some of the names you come across might be confusing. The word *design* is one such example since it usually implies what people see. However, in actuality, it means a fashion, or a formula to do or conceive something. Let's discuss it in the right context.
>
> We could have designed the level differently so that access to the door at the end of the pier would be challenging. Perhaps the light conditions are so poor that Clara needs some help to see an important clue. To make progress in the game, game design rules will define how the player will interact with the world. Perhaps it's enough for the player to click game objects in the world, while other times, it'd be better to have an inventory and a crafting system.
>
> Lastly, the visual design has nothing to do with the previous two design concepts. The cave walls could still be cave walls but instead of having a low-poly and stylized look, they might have looked ultra-realistic, where you could feel the stones were damp and covered with moss. Would this have added anything to the game and been fun? So, all these design principles are equally important and yet distinct.

The level, `Level-01.blend`, is available inside the `Blender Models.zip` file at the root of this book's GitHub repository. You'll most likely need it open so that you can use it as a reference when you are building the level in Godot.

We will start building the level by laying out different sections of it. Speaking of which, we must follow these steps to structure our first level:

1. Create a new scene and save it as `Level-01.tscn` inside the `Scenes` folder.

2. Place a **Spatial** node as root and rename it **Level-01**.

3. Create more **Spatial** nodes inside the root node with the following node names:

   - **Floor**

   - **Columns**

   - **Walls**

   - **Rails**

   - **SunkWalls**

   - **Props**

   - **Rocks**

   - **Sconces**

   - **Dock**

We'll be using these child **Spatial** nodes to store different parts of the level since we'll end up having a lot of parts in this level, despite it being very small. The following screenshot shows the node structure after our last effort:

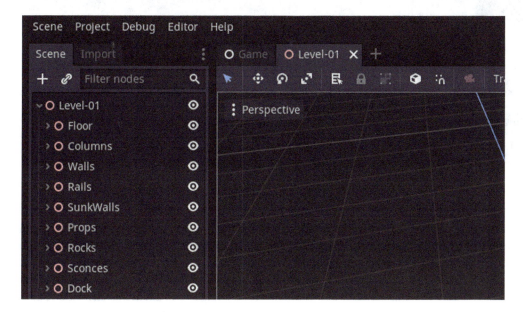

Figure 9.2 – Different structures for the level are grouped under many Spatial nodes

Inside these **Spatial** nodes, we'll place the relevant parts of the level. For example, the floor pieces will go inside the **Floor** node. We can put down our first asset easily by doing the following:

1. Highlight the **Floor** node in the **Scene** tree.
2. Press the chain icon at the top to instance another scene inside your highlighted node. Alternatively, you can press *Ctrl + Shift + A*.
3. Type `Floor_Standard` in the **Search** section of the pop-up screen.
4. Select `Floor_Standard.glb` from the **Matches** section, as shown in the following screenshot.

This will create an instance of `Floor_Standard.glb` inside the **Floor** node:

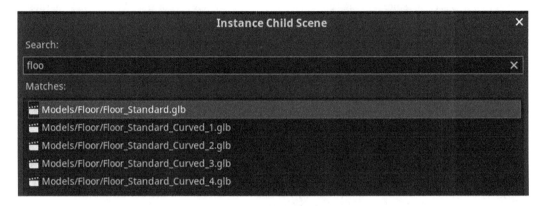

Figure 9.3 – You'll want to use the search area to filter out the unwanted matches

You may have noticed that although we wanted to inherit a scene that should normally have a `.tscn` file extension, instead, we instanced a glTF file. In *Chapter 7, Importing Blender Assets into Godot*, we learned how to create scenes out of glTF files. So, we could have done that and created a `Floor_Standard.tscn` scene, then instanced that scene inside the **Floor** node as well. We took a shortcut instead. Creating scenes is useful when you are going to add additional elements besides the model structure itself. We don't need additional elements for the floor, so it's alright to instance just its glTF version.

On the other hand, there will come a moment when we create our level when directly instancing glTF files won't cut it. For example, when we tackle lights and shadows in the next chapter, it will make much more sense to create a scene out of the sconce model and add a light object to the same scene. Hence, the sconce scene will take care of displaying a glTF model as well as holding a light object so that it can programmatically be turned on or off later. If you simply want to display models, but don't need anything more than that, instancing a glTF file is usually enough.

After you add the first piece, it will be automatically selected. If it's not, you can click the floor piece in the 3D view or highlight its node in the **Scene** tree. Once it's been selected, you'll see a gizmo at the center of the model that will let you move and rotate the piece around. The directions of your gizmo may look different if you have rotated your view. The following screenshot shows an example of what we expect to see:

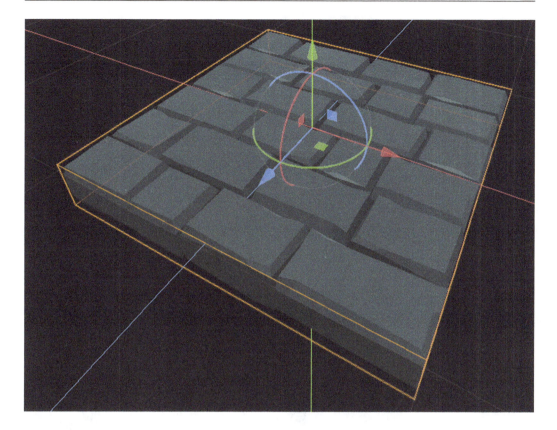

Figure 9.4 – The gizmo for moving and rotating an object

The floor plan we are trying to lay out consists of more standard floor pieces. So, an easy way to get extra pieces is to duplicate the existing pieces and move them aside, as follows:

1. Select the **Floor_Standard** node in the **Scene** tree.

2. Duplicate it by pressing *Ctrl + D*.

3. Move the new floor piece by dragging either the blue or the red axis in the gizmo.

This will add a new floor piece to the scene and move it around. We are intentionally ignoring the green (*Y*) axis since we don't want the floor to have any elevation at this point. However, for your games, you can design levels with different height zones and connect them with stairs.

Since our floor plan looks like a grid, it would be nice to have the floor pieces snap to each other. We can do this by moving the pieces in either direction on the *XZ* plane while limiting their movements to precise increments. To simulate this, delete the most recent floor piece you created, and then do the following:

1. Duplicate a new **Floor_Standard** node.
2. Hold down *Ctrl* and use either the X or Z gizmo arrow to move the piece two units.

Why did we move it by two units? Because the model is designed to fit in a grid that's 2 x 2 meters in size. You can open the relevant Blender file to observe the dimensions. We are not measuring things in Godot but it's still respecting the scale and unit aspects set in Blender. That's why we made sure the scale for the model was set to 1. If you need a reminder on this, we suggest that you read the *Applying rotation and scale* section in *Chapter 6*, *Exporting Blender Assets*.

After implementing the latest instructions for moving pieces with the snap feature on, you'll get the following output:

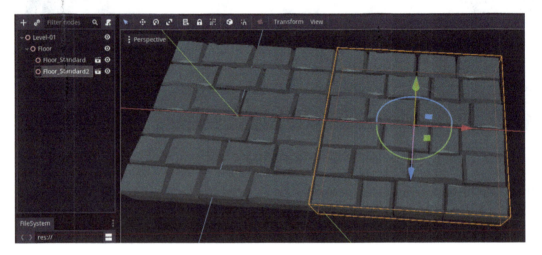

Figure 9.5 – The new floor piece is right next to the old one

All there is left to do at this point is duplicate enough floor pieces and move them around by using the snap feature. Also, you'll need to instance and place two new models inside the **Floor** node:

*   `Floor_Standard_Curved_1.glb`
*   `Floor_Standard_Curved_4.glb`

These curved floor tiles will accommodate curved walls, which means we can keep the architecture consistent. By duplicating enough floor tiles and adding the new curved pieces, and after moving the pieces around, we'll achieve the following output:

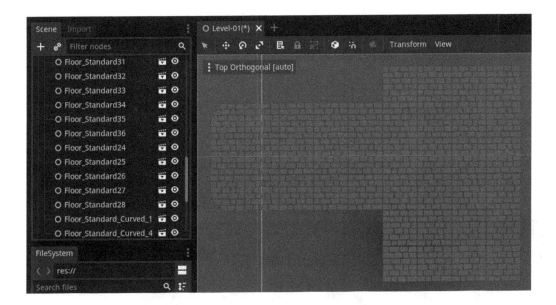

Figure 9.6 – With the two newly added types, the floor is ready

All the floor pieces are now under the **Floor** node in the scene, and this effort completes our task of constructing the floor. We'll follow a similar approach to lay the other parts of the cave under separate **Spatial** nodes.

Erecting the walls

The next order of business in constructing the level is putting up the wall sections. You can do so by instancing wall pieces under the **Walls** node, similar to the way you did for the floor pieces. As a substitute for providing you with very similar instructions, we'll use this section to highlight a few special cases you may come across.

For example, you'll eventually want to place wall pieces that will connect at a corner. So, you need to rotate one of the pieces around its *Y* axis by 90 degrees. You can do this either by using the gizmo or by typing the exact value in the **Inspector** panel under **Rotation Degrees** in the **Transform** section.

Another situation is with the wall that has a hole in it, which lets a bunch of twigs creep into the dock area. This is a detail you can see on the right-hand side of *Figure 9.1*. We suggest using `Wall_Hole.glb` for that particular section of the level. Similarly, `Curve.glb` should be placed over the curved floor pieces we have already established.

Although a door is technically not a wall, we could assume the arch and the door can get along with the other wall pieces. After all, they conceptually belong to the same structure. So, for that section, you can utilize the following pieces:

- `Wall_ArchRound_Overgrown.glb`

- `Arch_Round.glb`

- `Doors_RoundArch.glb`

Lastly, when you lay out all your wall pieces, you can duplicate them and pull them up two units on the *Y* axis. This will make the walls the same height as the arch and the door. Once you've done this, your floor should resemble what you can see in the following screenshot:

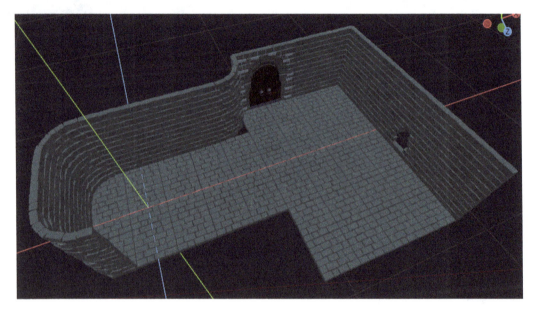

Figure 9.7 – The level is starting to look more like our reference picture

As you may have noticed there is a gap on the floor by the curved wall piece near the door. We'll fill that gap by cleverly placing two green plants soon. Otherwise, you'd have to prepare a floor piece for edge cases like that. Either way is fine and going back and forth between Blender and Godot to complete missing pieces is also part of the process.

Since we've been handling the walls, we can extend this effort by using additional wall pieces to simulate the section of the level that meets the seawater in the cave.

Sinking the walls

It seems the architect of this place went to great lengths to have stone bricks laid out to prevent mother nature from tarnishing what's under the floor. Smart!

To accomplish what the architect had in mind, you can utilize the standard wall pieces to create a curtain-like structure right where the floor is connecting with the water. In the end, when you place these pieces inside **SunkWalls** in your **Scene** tree, you'll be looking at what's shown in the following screenshot:

Figure 9.8 – The same wall pieces are used to prevent water from leaking under

The ebb and flow of the sea will now be kept at bay. Notice that we didn't want the sunken wall parts to go all the way around the floor. This is because you can always limit the camera angles to not show the back parts of the structure. It's a cheap way to keep the asset count low. However, if you want to give full freedom to the player so that they can rotate around the whole structure, you may want to change your level design to accommodate that. We'll be investigating camera settings in *Chapter 12, Interacting with the World Through Camera and Character Controllers*. For now, we still need to finish our level.

Placing the rocks

Since we are currently concerned about the parts near the water, let's add some rocks to the scene. In the Blender file for this level (Level-01.blend), you'll see individual rocks. They have been organized to give the illusion of a rock formation. It's perfectly fine to follow a similar approach and place specific rocks into your scene in Godot too, more specifically under the **Rocks** node.

However, there is an easier way. What if you exported the left and right rock formations as a single object from Blender? This is entirely possible, and that's why we have prepared two files for you:

- `RocksLeft.glb`
- `RocksRight.glb`

You can instance these two files and move the instances freely using the gizmo. This means you don't have to use the snap feature. Adjust the position of the rocks wherever you think is best.

Speaking of moving assets without using the snap feature, perhaps we can practice it a bit more. Since the floor looks empty, it's time we discuss complementary design elements such as props.

Distributing props

A prop is an object that serves as a support element. Props are also often called necessary clutter since they complete a décor. Otherwise, when things look too sterile, it's less pleasant to the eye and we start paying attention to repeating patterns or unnecessary details.

Instead, we want the person who's experiencing the scene to feel at ease. This is also a great way for designers to hide important elements in plain sight. To that end, we will use the following list of props and distribute these assets around the scene:

- `Barrel.glb`
- `Backpack.glb`
- `Bush_Round.glb`
- `Candles_1.glb` and `Candles_2.glb`
- `Cart.glb`
- `Crate.glb`
- `DeadTree_3.glb`
- `Flag_Wall.glb`
- `Pot1.glb`, `Pot2.glb`, `Pot3.glb`, and their broken versions
- `Statue_Stag.glb`

Once you've finished moving the props, your scene will look as follows:

Figure 9.9 – The props have been distributed all over the dock

While you are at it, you may as well instance `Column_Round.glb`, make two more copies, and place them under the **Columns** node. Also, `Rail_Corner.glb` and `Rail_Straight.glb` could be placed along the edge and near the stag statue. You don't have to be pixel-perfect with these objects, but if you want to be precise, you can use `Level-01.blend` for reference.

Finishing the rest of the level

To finish off the level, we need to place the sconces and construct a pier. These assets are no different than the other ones you have instanced and moved around the level.

However, placing the dock pieces may throw you off a bit as far as positioning goes. You may find that the stairs piece looks slightly off dimension-wise. Sometimes, assets are designed to be generic, while other times, assets will be designed so that they can fit or connect with the other models seamlessly. Regardless, since it's possible to adjust the final position in Godot, we can recover from these minor issues.

To simulate how we dealt with this issue, we'll give you the **Translation** values we used for the positions of both pieces:

- **Dock_Long**: `4,-1,5.5`
- **Dock_Stairs**: `4,-1.5,8.9`

Your values will most likely be different since you were undoubtedly moving your level pieces in directions that felt natural to you. If your numbers don't match our example, don't worry. We would like to point out the relative difference between the two structures. You'll also most likely have one number that's the same in one of the axes, either **X** or **Z**. Also, an educated guess on our end, your **Y** for the stairs will be 0.5 lower. This should result in a pier structure that looks like it was designed as one piece. If you want to have a taller pier, then you can create a copy of the stairs and move it accordingly. That's the benefit of having separate pieces.

We suggest that you add the boat model under the **Docks** node in the **Scene** tree at this point since it could be considered as part of the docks area. This concludes the construction of our level. It should look as follows:

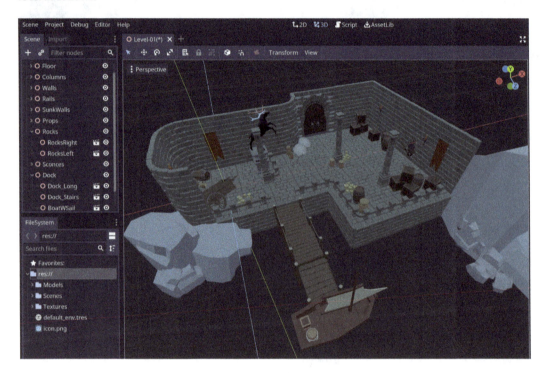

Figure 9.10 – The level has been reconstructed in Godot

Despite our claim that the level's construction is finished, you may have noticed that there are a few odd looking things. We have a dock area with no water – and what are those ugly round things doing by the door? We'll find out how we can remedy all this in the next section.

Constructing the missing materials

When we were placing the props, we covered the gap near the door by placing a bush prop (this can be seen in *Figure 9.10*). However, there is something awkward about those bushes. Similarly, the arch over the door has some weird-looking dangling things over the stone bricks. They should be showing greenery and leaves but all we have is a bland, gray surface. We'll fix these issues in this section.

In addition, while it made sense to export individual models from Blender and place them in a Godot scene, it didn't make sense to export the water body. Even in Blender, that object was a plane that has been applied a shader that mimicked water. We'll recreate that effect in Godot.

Fixing the leaves

First, let's describe what the problem is with the gray leaves. All the other models seem to have their materials displayed properly. Despite all intentions and efforts, certain things are never fully transferred between applications. This is the case with the leaves. We need to get a bit technical for a more thorough answer though.

How would you go about designing a leaf in 3D? Since a leaf has so many details around its edges, it's hard to display that much detail without using enough vertices. To be conservative, you can use an object with the least number of vertices and apply a transparent leaf texture to this basic object. The following screenshot shows an application of this method:

Figure 9.11 – A transparent file is used as a texture for a rectangle shape

The preceding screenshot shows a very simple shader. The alpha value of the texture is attached to the **Alpha** socket of the shader. Also, **Blend Mode** under **Settings** for the material is set to **Alpha Clip**. This means that the alpha parts of the texture will be clipped out of the result. We need to do the equivalent of this in Godot.

Unfortunately, Godot doesn't automatically understand and turn on transparency for imported materials. We'll have to do some manual work to display the leaves correctly. Luckily, this is also going to get you familiarized with the materials and their settings in the **Inspector** panel.

Let's start by finding the material for the bushes. The `Models` folder is structured in a way to keep distinct models inside individual folders. Hence, expand the `Bush` folder in **FileSystem** panel and double-click the `Texture_Leaves.material` item. This will populate the **Inspector** panel with this material's properties. There is a lot to look at, but we only need to tweak a few things:

1. Expand the **Flags** section.

2. Turn the **Transparent** setting on.

3. Expand the **Albedo** section.

4. Drag and drop `Leaf_Texture.png` from the `Textures` folder into the **Texture** field. As an alternative, you can click the **Texture** field and **Load** the necessary file.

As you may have noticed, the texture for the material was missing, so there was no chance for the bushes to display anything. Second of all, by turning the transparency on in the flags, we are asking Godot to respect the transparent parts of the texture file. You can switch it on and off to see the difference if you like. In the end, our scene will look as follows:

Figure 9.12 – Our bushes are starting to look healthier again

You can do the same thing for the arch model, which can be found in the `Architecture` folder inside the `Models` folder. This may look like you are repeating yourself, and you are right about this. Since we are keeping separate models that use the same Blender material inside their relevant folders, the materials are duplicated as well. A detailed discussion about this was provided in the *Deciding what to do with materials* section of *Chapter 7, Importing Blender Assets into Godot*. Since this is an organizational issue, we leave the decision to you, but you now know how to enable transparency in materials.

Another missing piece in our material puzzle is the water object. We intentionally omitted the export for that area. To most game developers out there, writing shader code is entering dangerous waters. Nevertheless, that's exactly what we'll do. Hopefully, you'll see that there is nothing to fear.

Creating the water

How do you model a body of water? The answer is not simple, and it even is a bit philosophical. The following is a homage to Bruce Lee's famous philosophical quote on martial arts, which uses water as an analogy:

> *"... Be formless, shapeless, like water.*
>
> *You put water into a cup, it becomes the cup.*
>
> *You put water into a bottle, it becomes the bottle. ..."*

It's hard to imagine what vertices we should create and organize for water in Blender or Godot. Instead, we give qualities of water such as reflection, refraction, undulation, and murkiness to simple objects, such as a plane or a cube.

Thus, for this effort, we usually rely on shaders instead of a 3D model. In this section, we are going to write a very simple water shader. In the end, you can either use the shader from our example or find another example on the internet. After all, there are a lot of examples out there, since creating a decent water shader usually depends on your use case, and one solution sometimes doesn't fit all.

Let's start by creating a water object:

1. Place a **MeshInstance** node under the **Dock** node and rename it **Water**.

2. For this new object, assign a **PlaneMesh** to its **Mesh** property in the **Inspector** panel.

3. Click this **PlaneMesh** to expand its properties, and fill in the following values:

    I.   `20` for both **x** and **y** in **Size**.

    II. `20` for both **Subdivide Width** and **Subdivide Height**.

We'll explain what these numbers mean soon, but here is what your **Inspector** panel should look like:

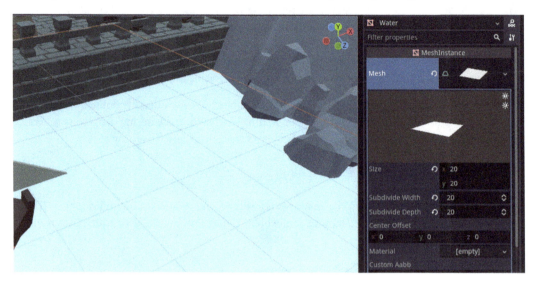

Figure 9.13 – A rather gray-looking body of water so far

The preceding screenshot shows the properties of a **PlaneMesh** in the **Inspector** panel. We have chosen a size that made sense as far as the level's dimensions are concerned. Using the gizmo, as you did for moving other objects, position the water object where it makes sense concerning the dock area and the overall scene. Once we have written our shader to make this gray object look like water, you may also want to adjust its **Y** position too.

Additionally, perhaps coincidentally, we chose **20** as the subdivision value. You can divide the plane into finer pieces if you want, but a value such as 20 will introduce enough vertices. So, yes, you have effectively created vertices in Godot as opposed to doing so in Blender.

We are now ready to change the look of this gray plane. For this, we'll create a material for it:

1. Right-click the res:// item in **FileSystem** and choose **New Folder**.
2. Type Materials and confirm.
3. Right-click the Materials folder in **FilcSystem** and choose **New Resource**.
4. Search for **ShaderMaterial** and confirm.
5. Save it as Water.tres in the upcoming **Save Resource As** screen.

Normally, a newly created item will be displayed in the **Inspector** panel, but if it doesn't or if you brought another object's properties to the **Inspector** panel, find Water.tres in **FileSystem** and double-click it. You'll see a barebones material with a white sphere as a preview in the **Inspector** panel. It needs a shader to get more water-like visual qualities. This is how you can create it:

1. Right-click the `Materials` folder in **FileSystem** and choose **New Resource**.

2. Search for **Shader** and confirm.

3. Save it as `cave-water.tres` in the upcoming **Save Resource As** screen.

In *Chapter 2*, *Building Materials and Shaders*, we discussed the relationship between shaders and materials, and how they go hand in hand. That was done in Blender, but the concept is universal. Hence, we've created a material and a shader in Godot. Now, we must associate the two:

1. Bring up the `Water.tres` file's properties to the **Inspector** panel.

2. Drag and drop `cave-water.tres` into the **Shader** property in the **Inspector** panel.

The water material has now been assigned an empty shader. We'll explain the shader code after you complete the following steps:

1. Double-click `cave-water.tres` in **FileSystem**.

2. Type the following code in the newly expanded **Shader** panel:

```
shader_type spatial;

uniform sampler2D wave_pattern;
uniform vec4 color:hint_color = vec4(0.19, 0.71, 0.82,
0.44);
uniform float height_factor:hint_range(0,1.0) = 0.1;

void vertex(){
    vec4 wave = texture(wave_pattern, UV);
    float displacement = sin(VERTEX.x * wave.x * TIME)
        + cos(VERTEX.z * wave.z *  TIME);

    VERTEX.y += displacement * height_factor;
}

void fragment(){
    ALBEDO = color.rgb;
    ALPHA = color.a;
}
```

The shader code we have written exposes a few options to the **Inspector** panel, starting with the lines that start with the uniform statement. This is so that you can modify the material's properties, just like you were able to change the settings of the leaf material earlier in the *Fixing the leaves* section. That one was a very elaborate shader with lots of options. Ours is a very simple shader with only three parameters:

- A wave pattern for creating randomness
- A color for the water (by default, this is a light blue color)
- A height factor to control the motion of the waves (by default, this is 0.1)

Two of the properties have their default values. We'll show you what you can pick for the wave pattern later in this section, but first, let's explain the general idea behind all this since this might be the first time you are writing shader code.

> **Built-in Godot shader functions**
>
> The two functions, vertex and fragment, are built-in shader functions. The former controls what each vertex will do, while the latter takes care of how the overall object will look. Godot has more default functions; we've provided a link in the *Further reading* section for you to discover.

Since the fragment function looks simple enough, we'll cover that one first. One of the properties we exposed, color, will be used in this function so that we can paint the object with the color we want. Consequently, we are taking the red, green, and blue channels of the input color and applying them to the ALBEDO property of the shader. Albedo is a scientific term for color. In some applications, it's also referred to as **Diffuse** or **Base Color**, such as in Blender.

Naturally, we would like to have some translucent qualities for our water object. For that, we are using the input color's alpha channel and binding it to the ALPHA property of the shader. It's a simple but effective way to create transparency. Speaking of which, if you comment out the vertex function, you should still be able to see the transparency because each function is responsible for one major aspect. However, they complement each other when used together. So, it's now the vertex function's turn.

It would be nice to have the body of water move up and down a bit. That's the reason why we have introduced more vertices to the plane mesh by subdividing it. The vertex function will take each vertex and change its y value to create an up and down motion. The last line in the function is responsible for that. How much should each vertex change though? Well, that depends on your use case. However, we came up with a displacement value that seemed appropriate and yet exciting enough to simulate a somewhat calm water feature in this cave.

While calculating displacement, we are using a texture and sampling some of its values. It'll bring randomness to the way the vertices will move. To that end, we are combining the x and z values of each vertex with the x and z values of the incoming texture (wave). You could alter a combination of some of those properties and still get a similar result. Perhaps what's more important is the use of the built-in TIME property, which is telling the GPU to change the result with each millisecond passed. Remove TIME from the equation and everything will be displaced once and sit still.

Finally, we regulate the intensity of the displacement with a height factor that can be adjusted in the material settings. This concludes our water shader. The shader and material have already been connected, but we have yet to tell the **Water** node which material it should use. To do so, follow these steps:

1. Select the **Water** node in the **Scene** tree.

2. Expand the **Material** section in the **Inspector** panel. You'll see a slot with a label of **0**.

3. Drag Water.tres from **FileSystem** to the **0** slot.

Voila! The dock should now have a water object that's modulating over time. Move and zoom your viewport camera in to get closer to the sunk walls to notice the alpha effect too. This is looking nice already, but we can take this a step further by applying the shader a noise texture, which will add more variation to the way the vertices fluctuate:

1. Expand the **Shader Param** section in the material's settings in the **Inspector** panel.

2. Attach a **New NoiseTexture** for the **Wave Pattern** property.

3. Expand this new texture and attach a **New OpenSimplexNoise** to its **Noise** property.

This will add more randomness to the way the vertices are displaced. When you are done with all the code bits and tweakings, your **Inspector** panel should look as follows:

Figure 9.14 – Notice how the water is transparent and wavy along the sunk walls

It's possible to fuss with the values of the noise to create more drastic effects, but we leave that to you. By controlling the height factor and color, you can simulate calmer or stormier water conditions as well. With that, you have created an important missing feature.

About keeping the shader separate

While creating the water material, you could have used an in-memory shader for the material using the dropdown in the **Inspector** panel. Most Godot features usually start and stay this way, but we followed a different approach by creating a resource first and then assigning it later. Thanks to this method, you can create different water shaders and swap them as you need them.

With that, we have taken care of placing all the necessary elements and even completing missing parts, such as fixing and/or creating new materials. However, while creating the layout, did it feel like you were duplicating and moving so many of the same assets, especially with the wall and floor pieces? We bet it did! So, let's present a very helpful Godot tool with which you can lay things out easily if your layout is grid-based.

Laying models on a grid

The main difference between placing objects such as candles, pots, and barrels, short props, and floor and wall pieces is that you can distribute the former objects willy-nilly. They don't have to follow a pattern, whereas the floor and wall pieces must snap to each other. This kind of structure is also referred to as a grid.

To speed things up, we even chose to duplicate an existing piece instead of instancing a fresh one because when you create a new instance, it'd start at the scene origin, and you'd have to move this new piece near your current area. You can even select multiple tiles in a row, duplicate them, and snap these next to the old batch. Despite all these shortcuts, since all this sounds formulaic, perhaps there should be a better tool. **GridMap** to the rescue!

If you have used Godot for 2D, you may already be familiar with the **TileMap** node. **GridMap** is the same except it works in 3D. Thus, whereas **TileMap** will let you add sprites to your scene, **GridMap** will use meshes. For those of you who have never used a **TileMap** node, both of these mechanisms in Godot are responsible for using a set of tiles or meshes.

Benefits over manual placement

The **GridMap** solution we are offering is not just for you to expedite the creation of your levels. Since the pieces are repeating, the GPU will optimize the rendering of said pieces and you'll get higher frame rates. This is usually a very sought-after result among game developers, particularly when your levels grow and the number of objects you use in a scene starts to matter.

In this section, we'll present the general settings of a **GridMap** node. Although this node depends on **MeshLibrary** to do its job, it makes sense to understand the individual settings at this point than mixing both. We'll learn how to create and utilize **MeshLibrary** in the *Taking advantage of MeshLibrary* section.

To conserve and compare what we have done so far, we'll take things a bit slowly:

1. Save `Level-01.tscn` as `Level-01-Gridmap.tscn`. The root node could still stay as **Level-01**.

2. Add a **GridMap** node and rename it **FloorGridMap**. You can drag this new node and make it the first child right above the **Floor** node if you wish.

3. Turn off the **Floor** node by pressing the eye icon.

The last set of actions will introduce a **GridMap** node to the scene. It's empty for now but we'll fill it with the floor pieces when we get to know mesh libraries. Your scene will look as follows:

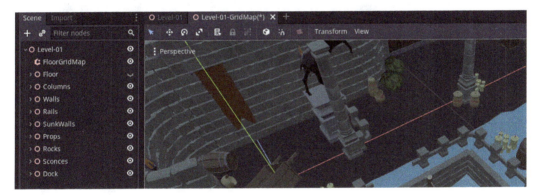

Figure 9.15 – The missing floor pieces will soon be introduced with GridMap

Although we're missing a mesh library, we have a **GridMap** node for which we can look at properties in the **Inspector** panel. We suggest that you select **FloorGridMap** now and read along. The information we'll present here will lay the foundation for you to choose the settings of the future grids you'll use.

GridMap works with a cell concept. A cell is a volume in which one of the meshes will fit. If you expand the **Cell** section in the **Inspector** panel for the **FloorGridMap** node, you'll see that we have a value of 2 across the board for a cell. Fortunately, our floor pieces are 2 x 2 x 2 meters as well. So, we don't need to change those values in our case. In your future projects, you may have to match these values to your models' dimensions.

We'll ignore the **Octant Size** setting in our efforts since it's for more advanced cases where you can further increase optimization. What's much more important perhaps is the three on/off switches for centering the meshes inside a cell on either axis. We'll make use of this very soon, but the following screenshot should help you see what we have been discussing so far:

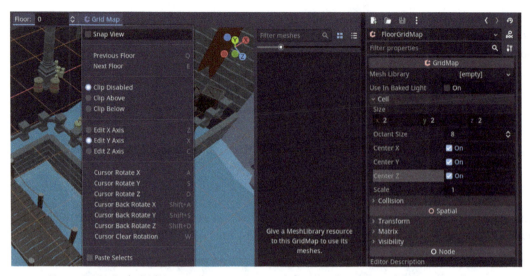

Figure 9.16 – Each GridMap can have settings to define the dimension of the pieces it'll use

The preceding screenshot also shows an expanded menu and its options when you click the **Grid Map** button at the top of the viewport. Out of those options, **Cursor Rotate Y** with the *S* shortcut will probably be the one you'll use the most. The floor pieces we laid out earlier in the *Creating the cave* section all follow the same direction. We tried to cover the floor with props to break the sameness but rotating a floor piece 180 degrees around the **Y** axis would be another solution.

Now that the theoretical knowledge has been established, let's move on to practical applications of using **GridMap**. In the next section, we'll create a mesh library that we'll use in tandem with our **FloorGridMap** to fill in the missing floor pieces.

Taking advantage of MeshLibrary

When you clicked **FloorGridMap** to investigate its properties, the Godot interface changed slightly, and it informed you that you should assign a **MeshLibrary** since, without one, a **GridMap** is ineffective. In this section, we'll show you what goes into creating a **MeshLibrary**. We'll also talk about possible challenges you might face, not technically, but workflow-wise.

There are two ways to create a **MeshLibrary**. We'll show you the most common way since the other method involves keeping meshes separately in the filesystem, and our project has not been set up to accommodate that scenario. Without further ado, this is how you create a mesh library:

1. Start a new scene and save it as `Floor-MeshLibrary.tscn` in `Miscellaneous`.

2. Choose a **Spatial** node as its root.

3. Instance **Floor_Standard** under the **Spatial** node in the **Scene** panel.

4. Click the **Scene** button in Godot's top menu.

5. Expand **Convert To** and choose **MeshLibrary**.

6. Save your mesh library as `Floor-MeshLibrary.tres` in `Miscellaneous`.

If you drag and drop the floor piece directly into the viewport, it will be placed somewhere in the scene while considering the perspective of where your mouse cursor was. The floor may, for example, look tiny because it will be far away from you. Zeroing the position should put the object in the center of the world and bring it closer. If you dropped the piece into the **Scene** tree instead, you won't have this problem.

The following screenshot shows the state right before Godot converts your scene into a mesh library:

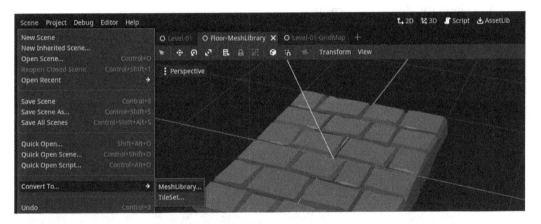

Figure 9.17 – We are converting a scene into a mesh library

Now that we have a floor piece in the library, we can add one more model to it. The goal here is to pile up items that have similar dimensions. This may sound confusing, but let's add the curved wall. Why? Because although a wall is normally thinner and taller, if you think of the volume the curved wall occupies, it's no different than a floor piece. Its base is of similar dimensions.

So, assuming `Floor-MeshLibrary.tscn` is still open, here is how you can introduce another model to the same library:

1. Find the `Curve.glb` wall piece in **FileSystem**.

2. Drag and drop it over **Spatial**.

3. Convert your scene into a **MeshLibrary** again and overwrite the existing file in `Miscellaneous`.

This operation will add the newly introduced piece alongside the old floor piece and update the mesh library. Thus, an easy way to create a mesh library is to start a new scene, add as many models as you want, and turn this scene full of models into a mesh library.

We haven't concerned ourselves with where the pieces will go yet. We've just been selecting separate pieces as candidates to decorate a grid. Now, let's associate the mesh library with **FloorGridMap** and start laying some models.

Using a mesh library with a grid map

So far, we have been preparing a mesh library to be used by **FloorGridMap**. We have two pieces inside this library. We'll use the floor piece first, and then see if it makes sense to use the curved piece.

For a **GridMap** to work, you need to fill its **Mesh Library** property in the **Inspector** panel. Let's take care of this first:

1. Select **FloorGridMap** in the **Scene** tree.
2. Drag and drop `Floor-MeshLibrary.tres` from `Miscellaneous` into the relevant field in the **Inspector** panel.

This will display all the available models as thumbnails in the reserved **GridMap** interface, as shown here:

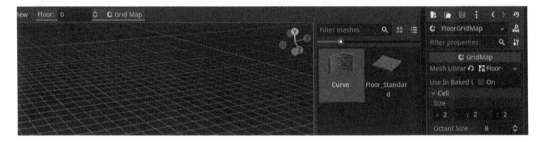

Figure 9.18 – The mesh library can now be used by FloorGridMap

All there is left to do is click one of those thumbnails – for example, **Floor_Standard** – and move your mouse over the viewport. You should see a preview of the selected model under your cursor. If you click where you can see the preview, you'll make it permanent. Try this a few times.

Isn't this a lot easier than laying out all the floor tiles by yourself? But wait a minute – you'll most likely notice that something looks slightly odd. It's as if the floor pieces are not quite where they are supposed to be. They snap to each other, but they don't seem to quite respect the old coordinates. They are either elevated, penetrating wall pieces, or situated off the walls.

This is something you'll regularly come across when you work with grid maps. The solution is easy, but keep in mind that this is not exactly a problem either. It depends on the origin points you set for your models. So, yes, the origin points are something you may have to deal with even after you have exported your models. As a result, you can either fix your origin points by going back inside Blender and re-exporting your models or use some of the options available to you in the **Inspector** panel.

For now, let's try to turn the following **Cell** settings on and off:

- **Center X**
- **Center Y**
- **Center Z**

There is no set formula for whether these properties should be on or off. It depends on the models that are used in a mesh library. For example, the **Curve** piece in the mesh library has its origin point in one of the corners, whereas the floor piece has it, geometrically speaking, in the middle. Since there is only one **Cell** setting for the whole grid map, you must have a standard way of dealing with all the models of a mesh library. So, it's not just about piling up a whole bunch of models haphazardly – it's about storing them in a way that respects cells, hence a grid structure.

To visualize what we are talking about, you can try to place a **Curve** piece from the mesh library onto the scene. You'll notice that you'll have to reset the center settings but that this will also reset the floor pieces back to their controversial positions. Therefore, this is something you've got to plan for and make sure your objects share similar origin points, as well as similar dimensions.

Clearing a cell

You already know that clicking with the left button of your mouse will place the previewed item from the mesh library. If you need to remove an existing cell from your scene, you can right-click it and move your mouse around. If you happen to have the same model in preview mode, removing the cell from the scene but not moving your cursor anywhere else may give the impression that you didn't remove anything. So, remember to wiggle your mouse after you clear a cell.

The necessity of using multiple grid maps

Either for the reason that the dimensions of your models will be different, or the origin points won't necessarily align, you'll eventually notice that you're going to need to use different grid maps in your scene. Since each grid map can have separate **Cell** settings, it's entirely possible to use the same mesh library among all these grid maps.

In this scenario, you'll have the convenience of creating one mesh library to store all similar items – for example, all the architectural elements – but only use some of the models for the right grid map. This beats the hard work of creating individual mesh libraries.

Wrapping up

Using grid maps is a convenient way to distribute objects that follow a pattern. The decision to use it is sometimes an organic process. Most people often start building their level by individually placing items. This is usually when they aren't using an already existing level design software. So, the process of creating a level happens while you are moving stuff around in a natural way, similar to moving furniture around instead of using a floor planner.

Thus, either you decide early on or feel the need to switch to it, using grid maps will make your life easier. That being said, grid maps and mesh libraries are full of bugs in the current version of Godot. For example, adding new models to your mesh library scene, then exporting it as a library, won't always update the existing library with new models. Sometimes, the earlier items within a library will be swapped with the newer models. So, it's quite inconsistent. Hopefully, the fourth version of Godot will eradicate all these problems.

We wanted to be comprehensive about different ways to create your levels. So, it felt necessary to introduce the **GridMap** node, however broken it might be. This way, when the community gets this tool implemented bug-free in the future, you know that such a convenient tool will be available and useful.

Summary

This chapter was the first out of many chapters that will help you build a game. To kick things off, we tackled the level design aspect of the game.

This effort involved placing many elements that make up the environment Clara will experience. For structures that are next to each other, you learned how to take advantage of the snapping feature, but you can also decorate your scene carefree if you wish, in the case of distributing props. In the end, you had a clean scene structure with objects grouped under the relevant nodes in the **Scene** tree.

Along the way, you noticed that some of the materials were either misconfigured or simply missing. To fix these issues, you had to dive deeper into the **Inspector** settings for materials with which you remedied the transparency issue. Furthermore, you wrote a shader in Godot to simulate a body of water.

Considering what you have learned so far and the likelihood that you might be designing more levels that have grid patterns, we presented Godot's **GridMap** node. To be able to use this handy tool, you also learned how to create a **MeshLibrary**. Despite its benefits, this last method is broken at the moment, but it's something you can employ in future versions of Godot.

With that, the level is complete to the point that you can start adding a few more elements down the road as you need them. Despite that, everything looks a bit bland. In the next chapter, we'll learn how to make the level look fancier with lights, shadows, and environmental effects.

Further reading

Level design doesn't always involve placing physical elements inside the game world. Sometimes, it means enticing sound design, hiding cute or interesting lore elements pertinent to the world and story, and adding non-player characters your players can relate to or simply hate. There is a whole layer of psychological factors to designing good levels so that you can evoke the emotions you desire in your players. If you want to elevate your knowledge in this domain, you are going to have to examine resources that are not necessarily game engine-specific. So, broaden your horizons! Here are a few resources that will get you started:

- `https://www.worldofleveldesign.com`
- `https://www.pluralsight.com/courses/fundamentals-professional-level-design`
- `https://www.cgmasteracademy.com/courses/46-level-design-for-games/`
- `https://www.edx.org/course/introduction-to-level-design-2`

You had to write a water shader in this chapter. Working with shaders is often described as the least entertaining or the most confusing experience among game developers. We'll give you two links so that you can familiarize yourself with this topic. The former is the official Godot documentation, which should help you produce more direct results in your projects, while the latter should be useful for more long-term needs:

- `https://docs.godotengine.org/en/stable/tutorials/shaders/`
- `https://thebookofshaders.com/`

10
Making Things Look Better with Lights and Shadows

We have a simple and clean-looking level design, but it could use a good makeover. For example, the sconces on the walls and the candles on the floor are just sitting there without adding much interest to the scene. Also, there is the slight issue of having this level as an underground environment since this is a cave. We must find a way to simulate the light from the exterior since Clara sailed her boat in. Overall, we will have the level be lit just enough for the players to perceive things.

In this chapter, we'll introduce lights and shadows to our workflow so that our scene looks visually appealing. We covered lights earlier in *Chapter 4, Adjusting Cameras and Lights*, but we did that in the context of Blender. While generic concepts still apply, we'll have a chance to do things from a game development perspective this time instead of taking an artistic render in Blender.

Shadows are not automatically available in Godot. Therefore, we'll show you how to turn them on and discover some of the shadow settings that balance quality and performance. Besides placing light objects and enabling shadows, and altering their qualities, we will present a higher-level concept, creating a **WorldEnvironment**. This is also referred to as post-processing and it's a great tool to improve the look and feel of your scenes.

Although we'll be improving the level with each new addition of the topics we have listed so far, to tie this all together, we'll also tackle a somewhat advanced topic, **global illumination**, which will add a realistic touch to the scene.

We have many steps to take before we will have created a handsome-looking level. In this chapter, we will cover the following topics:

- Adding different types of light
- Enabling and adjusting shadows
- Creating post-processing effects
- Using global illumination

Even though the purpose of this chapter is to understand how the lighting system works, we'll introduce a few complementary Godot topics along the way.

By the end of this chapter, you'll be able to utilize lights and enable shadows, as well as to take advantage of global illumination and post-processing effects that will further enhance the atmosphere in the level.

Technical requirements

We'll add and change things from where we left off. You have two options at this point – you can either keep working on your copy from the previous chapter or use the `Finish` folder mentioned in *Chapter 9*, *Designing the Level*, which is available in this book's GitHub repository: `https://github.com/PacktPublishing/Game-Development-with-Blender-and-Godot`.

Adding different types of light

In *Chapter 4*, *Adjusting Cameras and Lights*, we discussed how different types of light worked – more importantly, the kind of effect they bring to a scene. In this chapter, we'll revisit the same topic but pursue the effort in the context of Godot.

Blender uses four light types: **Sun**, **Point**, **Spot**, and **Area**. However, Godot has only three lights, as follows:

- **DirectionalLight**: This is the equivalent of the **Sun** light in Blender. We stated directionality in the **Sun** light's description. The angle of this light type is the most important since it's an infinitely distant light source, so all its rays are considered to flow parallel to each other. So, in Godot, this concept is part of the node's name, hence making it easier to remember.

  We'll not be using this type of light in our scene since it's an indoor environment. Despite that, it may still be tempting to utilize it to give an overall light effect, but this light source would overwhelm the whole scene. We need something else that can be fine-tuned as we go. Therefore, we'll focus on the two other light types.

- **OmniLight**: This is what the **Point** light is in Blender. Lightbulbs and, yes, the sconces on the walls, are the right kind of objects for which this type of light is good. As a reminder, omni means in every direction.

- **SpotLight**: This one is self-evident – it's the **Spot** light in Blender. It's good for simulating car lights, flashlights, and any other light source that has a beam-like quality. We'll be using this light to simulate the exterior light creeping into the cave.

So, where is the **Area** light in Godot? It simply doesn't exist. There are different mechanisms in Godot that you can use to simulate the effect of an **Area** light in Blender. Often, this kind of light is for mimicking the light coming in from a window, and it can be simulated with emissive materials.

Speaking of using different types of light, let's start by lighting those candles.

Lighting candles

For this type of exercise, the **OmniLight** type is the right choice, but how many are we supposed to have? If you look closely at the candle model, you'll see that the model is composed of multiple candles; some short, some tall. Does it make sense to place one **OmniLight** above each wick? It's entirely possible but it's also an artistic decision to make, and we leave it to you.

In our case, we'll assume that the overall light coming from this object could be reduced to a point over the candles' wicks. Thus, it's perfectly fine to place one **OmniLight** for the whole model. It's time to demonstrate how this can be done:

1. Double-click `Candles_1.glb` in **FileSystem** to create a **New Inherited** scene. Save it as `Candles_1.tscn` in its original folder (`Models/Candles/`).

2. Add an **OmniLight** to the **Scene** tree.

3. Adjust its **Y** position to `0.8`, for example, so that it's slightly over the wicks.

This will place a point light in your candle scene. Right now, with the default settings, it's hard to see the impact. If you get close to the light object and adjust your camera angle so that you no longer see the horizon and the sky, you can get a better view. Perhaps turn the visibility of **OmniLight** on and off in the **Scene** tree to see the light's contribution.

We'll leave most of the settings in the **Inspector** panel untouched for now, but you can change the light's color to, for example, `d6d58e`. This can be found under the **Light** section in the **Inspector** panel. The result is as follows:

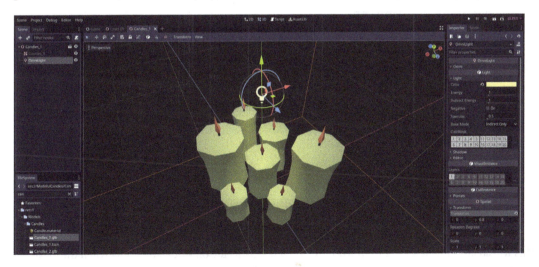

Figure 10.1 – An OmniLight with a yellow color right over the candles

Let's take a moment to discuss why we've added a light to a scene that we are constructing out of a model instead of directly adding it to the level scene. After all, we already had several **Spatial** nodes to hold items that were similar, such as walls, columns, and more. We could have created a **Spatial** node called *Lights* and stuck a bunch of **OmniLight** nodes in there.

By introducing a light node to the model scene instead of the main level, you can utilize this candle scene in other scenes as well. Hence, you don't have to create more light objects and position them over all the candles in the level. When you are decorating a level, and when you need candles, they will arrive as a full-service package.

> **Overcoming the jagged edges**
>
> After adding lights to the scene, you may notice that some objects look jagged around the edges since the details are popping up more. To eliminate this, you can turn on the anti-aliasing setting. The hard edges will be smoothed out, objects will blend more seamlessly, and everything will look easier on the eyes. To enable it, set the **Msaa** value to **2x**. This setting can be found under the **Quality** subsection of **Rendering** in **Project Settings**.

So far, so good, but will the lights always be on? It seems so, for now. Let's see how we can complete the full-service aspect of the candles by introducing a mechanism that will switch the lights off. To do this, we need to add a short script:

1. Right-click the root node (**Candle_1**) and choose **Attach Script**.

2. Keep everything the same in the upcoming pop-up screen, but change the path so that it shows /Models/Candles/Candles.gd.

3. Your script file should contain the following lines of code:

    ```
    extends Spatial


    export(bool) var is_lit = false setget switch


    func switch(condition):
        is_lit = condition


    func _process(_delta: float) -> void:
        $OmniLight.visible = is_lit
    ```

This will create a toggle state for **OmniLight**, and it'll start its life off by default. Only when the player, or you as a developer, change the value of is_lit will the light become visible again.

To test this without running the game, you can add the `tool` keyword at the beginning of the script and see your changes live while you are still working on the level. Observe how the light's visibility changes in the **Scene** panel when you toggle the state of `is_lit` in the **Inspector** panel.

We have another candle model, `Candles_2.glb`, that could also benefit from all this. Instead of starting from scratch, this is what we suggest you do:

1. In the **Candles_1** scene, right-click the **OmniLight** node in the **Scene** tree and choose **Copy**.

2. Create a scene out of `Candles_2.glb` and save it in its original folder.

3. Right-click the root node of this new scene and choose **Paste**.

4. Select the root node and attach `Candles.gd` to the **Script** property in the **Inspector** panel.

This will minimize the number of steps you have to take to add an **OmniLight**, position it, then write pretty much the same script for controlling it. Here, we are using the same script for both scenes since the node references in the scene are the same. After making our most recent changes, the Godot editor will look as follows:

Figure 10.2 – A new candle scene using the same script for switch functionality

Although we have been working on a smart way to add lights via attaching light objects to the candle models, we haven't made any changes to the level itself. We will discuss this next and share a few words about workflow improvements that you can make in your future projects.

Introducing candles to the level

In *Chapter 9*, *Designing the Level*, we instructed you to instance glTF files directly to the level, which kept the filesystem clean without creating redundant `.tscn` files. Otherwise, you'd have had one scene file per model with no purpose at all. A simple workflow such as only adding the models to a scene is often enough, especially in cases where you don't have prior knowledge of where your project is headed.

On the other hand, in certain cases, such as where you have candles and sconces, you will most likely have a light node beside a **MeshInstance** node, as well as a script attached to control the light's behavior. In that case, it pays off to convert the model into a scene and build up from there.

The **Scene** tree for the level still holds the raw candle models. In *Chapter 9*, *Designing the Level*, we used two types of candles but three models in total to decorate the level. It's perfectly alright to remove these models from the level so that you can instance the new candle scenes. You would have to reposition these new items though. So, we'll follow a different path to keep the position information:

1. Select **Candles_1** in the **Scene** tree.

2. Instance `Candles_1.tscn`, which will result in a nested node.

3. Drag this nested node out of its parent and make it a sibling of its parent.

By nesting the candle scene inside the old model instance, we are appropriating the position. If you added the candle scene directly into the **Props** node, you'd have to find the position of the model instance and apply it to the new item.

You can repeat this process for the other two candles, which will eventually double the number of visible candles in the level. That being said, our initial three candle model instances are no longer necessary, so you can delete them. Also, notice how a script icon appears in the **Scene** tree when you bring the candle scene versus keeping just the model itself. The following screenshot shows the result:

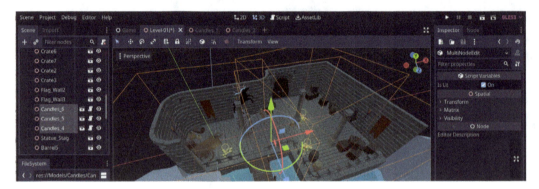

Figure 10.3 – The new candles in the Scene tree have script icons

The preceding screenshot shows not only the more advanced candles that have been added to the level but also the fact that you can turn these candles on and off via the **Is Lit** property in the **Inspector** panel. Similar to what you've done for the candles, you can continue practicing point lights by creating a scene out of the sconce model. In that case, the light object's position in the scene will most likely be higher since the model is taller, but the concept is the same. You can even bind the same script to the root of this sconce scene.

This creates a bit of a dilemma though. So far, we have kept everything related to candles in their own folder, with the script included. However, the light switch script is so generic that it could be used within any scene that has a similar structure. Although it's also possible to attach the `Candles.gd` script inside the `Candles` folder to a scene in a different folder, if you want to generalize things, you can move the script file into a separate `Scripts` folder at the root of the project.

This is one of many project management conundrums you'll face, so it's up to you how you want to go with it. We've decided to keep things as generic as possible. Hence, the `Finish` folder of this chapter will have both the candles and the sconce share the light script from the `Scripts` folder.

After swapping the sconce models with the sconce scenes, the level will have a bit more character, as shown here:

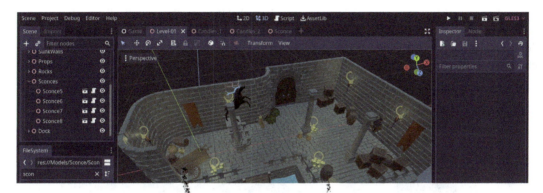

Figure 10.4 – Three candles and four sconces are illuminating the level

We have the basic lights covered, but we still don't have the kind of light effect you may see inside a cave. The idea is that Clara used an opening to enter this structure, so it makes sense to get some sunlight into the general area. We'll achieve this by using a **SpotLight** node.

Mimicking the sunlight

The narrative in our game is that the dock area Clara secured her boat to wasn't too far off from the entrance. Hence, it makes sense to get some sunlight from the exterior. An easy way to get an effect like this is to use a **SpotLight** node. Let's also discuss an alternative.

Using a **DirectionalLight** node seems tempting at first, but that would brighten the whole scene. Also, we want this cave to look as dark as possible, and only to be illuminated with artificial lights such as candles and sconces. To achieve both goals, you'd have to position planes over the level, pretending that they're the cave's ceiling, to block most of the light. So, since that kind of effort feels counter-productive, we'll try to light what we need instead of blocking the light.

Therefore, using a **SpotLight** node seems to be the best choice we have. We'll describe the process we used to place the light over the level so that it highlights the boat and a portion of the pier. Here we go:

1. Select the root node of the level (**Level-01**).

2. Add a **SpotLight** node and position it over the boat seven units or so in the **Y** direction.

3. Rotate it **-70** degrees in the **X** and **Y** directions (hint: use **Rotation Degrees** under **Transform** in the **Inspector** panel).

4. Change its color to d6d58e.

5. Expand the **Spot** section in the **Inspector** panel and set the following values:

   I.   Set **Range** to 20.

   II. Set **Angle** to 55.

We'll provide you with a screenshot right after we explain what the intention with the light's placement is and give a disclaimer about the screenshot itself. Since the default environment settings in your Godot project will result in a scene that's too bright for you to see the impact of what you are doing, we temporarily tweaked some settings to better highlight the contribution of the light you are working with.

We'll study environment effects in the *Creating post-processing effects* section after we finish exploring lights and shadows. For now, we still owe you an explanation about the settings of the **SpotLight** node. Even when you've been following a similar layout, the coordinates you have picked for your floor tiles might be so different that there is no easy way to ask you to place the light in a certain position. Hence, we are giving you a mix of general and precise directions. This is what we have got so far:

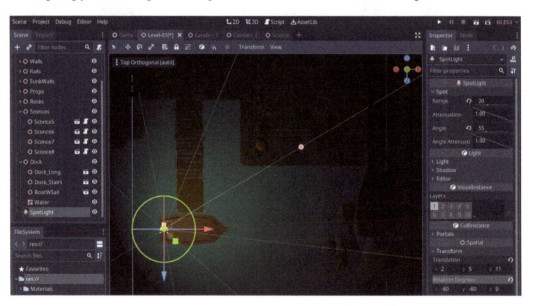

Figure 10.5 – The SpotLight node simulating the sun in the cave

The preceding screenshot shows the **SpotLight** node right above the boat's back. We chose the top-down view for you to see how far the light goes from this object. The **Range** and **Angle** properties you set in the **Inspector** panel will configure this light source so that it reaches far and wide enough to illuminate the entrance partially. Thus, if your layout necessitates different values so that you have an area lit just enough, as shown in the reference picture, you may have to alter the rotation and position values.

If you fancy, you could create another **SpotLight** node and alter its values as if there is a secondary opening in the rock formation that is letting more light through. Once you figure out the technical parts, it's up to you to push the envelope for an artistic result that pleases you.

So far, we've been analyzing different types of light and their impact on our level. With light, we usually expect shadows. These are not enabled by default, so we'll discover how to turn them on, as well as adjusting a few settings in the context of our project.

Enabling and adjusting shadows

In some situations, such as in stage arts, engineers work hard to illuminate parts of a stage with lights by casting their beams from so many angles that shadows can be eliminated. That's an extreme case. Normally, a shadow is something that occurs naturally when there is a nearby light source.

Despite this natural phenomenon, simulating shadows doesn't automatically happen in computer simulations just because there is a light object. The GPU has to know where the light is coming from and how strong it is. So, it can create an area, starting from the base of the object the light is turned to, and stretch this area out gradually in the opposite direction to the light by blending it into the surface the object is standing on. This is approximately how shadows are calculated and simulated by computers.

In Godot Engine, a light source is responsible for its own shadow. This means the shadow settings are part of a light object, but since the effort is resource-intensive, Godot has this property turned off by default. Let's look at an example and see how we can enable it:

1. Double-click the `Candles_1.tscn` item in **FileSystem** to open it.

2. Select the **OmniLight** node and expand its **Shadow** section in the **Inspector** panel.

3. Turn the **Enabled** property on.

The color of the shadow is irrelevant at this point, but it might be something you can tweak in your projects to get the dramatic effect you wish. At this point, we advise you to open the **Candles_2** and **Sconce** scenes to enable the shadow for the **OmniLight** nodes they have. When you save all these three files and go back to the **Level-01** scene, you should see something similar to the following:

Figure 10.6 – Let there be shadows, and shadows you shall have

Notice how enabling shadows elevates the experience overall. The column, the crates, and the other objects have started to come to life. There is one big missing piece in this picture, though: we still haven't enabled the shadows for the light source we are using to simulate the sun's effect. Go ahead and turn its shadow on; you'll make the pier pop up, as shown here:

Figure 10.7 – The pier and the boat look more realistic thanks to the sun's shadow effect

We are slowly improving the visual quality of the level. Our last effort introduced shadows. They are nice and all, but sometimes, they can also create a few defects. Now, let's talk about some of the settings you can find in the **Shadow** section of light nodes in the **Inspector** panel:

- **Bias**: Some names you come across in game development will sound technical, and won't always give you a quick idea about what they control. This one certainly sounds like one of those. In simple terms, this property controls where the shadows are going to start in comparison to an object's volume. A picture is worth a thousand words, so please refer to the following diagram to see what different **Bias** values will lead to:

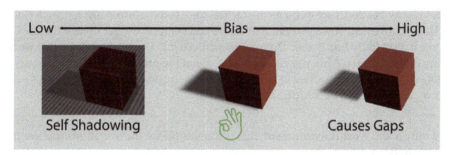

Figure 10.8 – Different bias values and their effects

- **Contact**: When you have a high **Bias** value, and it creates a gap between the shadow and the object (as shown in the preceding diagram), this property will try to fill in that gap.

So, if you happen to have visual glitches due to enabling shadows, which may result in shadows not always meeting an object or self-shadowing issues, as shown in the preceding diagram, we suggest you explore using a combination of the **Bias** and **Contact** properties for your lights.

The level is starting to look like there is more life to it, thanks to lights and shadows. Still, everything looks a bit too bright. If only we could dim the overall brightness… We certainly can, and that's what we are going to explore next.

Creating post-processing effects

Since we are pretending that Clara is visiting a cave that's got some human traffic in its past that led to having a pier built and sconces hung on the walls, it's only normal to expect some areas of it to be really dark. We have been placing lights and turning on shadows to improve the visual fidelity of our scene, but we are fighting against the environment; it's just too bright.

In this section, we'll study an interesting Godot node that will control the environment or world settings so that you have a much better hold on how your world looks. This kind of process is also referred to as post-processing since its effects are applied after the directly placed elements such as lights, shadows, reflections, and others have been processed. It comes with a lot of settings, and hopefully, this will be clearer after we explore some.

A node for everything

If you are coming from Unity, then the node system Godot uses might be confusing. In Unity, you attach scripts to game objects to add or control the behavior of systems. Nodes are analogous to scripts in Unity, but nodes are much more practical since you can also attach scripts to Godot nodes. This is convenient since you can nest nodes and compose bigger node structures. In Godot, you'll most likely find a node that will do a crucial job. One such node is what we are discussing in this chapter. Also, you can find more about the process behind using nodes in the *Godot's design philosophy* section at `https://docs.godotengine.org/en/3.4/getting_started/introduction/`.

Godot has a nifty node, **WorldEnvironment**, that is responsible for the overall atmosphere in your scenes. Although the node's name is quirky, introducing it to the level is no different than adding other nodes:

1. Open the `Level-01.tscn` scene.

2. Add a **WorldEnvironment** node to the **Scene** tree. For its **Environment** property, use `default_env.tres` from **FileSystem**.

3. Double-click `default_env.tres` in **FileSystem** to populate the **Inspector** panel with its properties.

Chances are nothing has changed, but we have effectively created a **WorldEnvironment** node and attached an environment resource to it. When you create a new Godot project, it comes with a default environment resource. Instead of creating a new resource, we are repurposing the default environment resource that's been sitting in the project folder all this time.

This opens up different possibilities for you. Your game may have different levels where you would like to have the visual clues support the characteristics of a particular level. In a situation like that, your project folder could store multiple environment resources and use them accordingly in the **WorldEnvironment** node.

Although the **WorldEnvironment** node's purpose may sound self-evident by its name, to fully take advantage of it, it would be best if you practice using its properties. You can do this by looking at the properties of the resource it's using. There are quite a few and we'll discover the ones that are relevant to our goal.

Background

This part of the environment's settings is responsible for simulating the background. Currently, the mode is set to **Sky**, so the background is painted as if there is a dark ground portion that goes out far enough to meet the sky. In this mode, you can further customize the properties of the sky you want to depict. We won't cover this since we are working with an indoor scene.

Thus, start by changing the mode to **Custom Color**. This will pick a black color by default, so the whole background of your scene will be pitch black. This will surely accentuate the candles and the sconces.

If you would like to use Godot Engine to take in-game renders of your models, then you can set the background to **Clear Color**, which will create a transparent color. We're not using it in our case since having a completely dark background suits our artistic needs better and also, the body of water looks a bit awkward with transparency underneath. We'd need another similarly sized dark plane under it to hide the effect of transparency.

Therefore, we'll stick with a custom background color. This will result in the following output:

Figure 10.9 – The cave is starting to look more ominous

Just a quick discussion about the **Ambient Light** section before we move on to **ToneMap**. The arched door seems to be hidden right now because there aren't enough lights in the scene. So, to remedy this, you could pick a lighter ambient color. However, this will make the overall scene brighter again, and you'll have some of the dark areas more lit. There is a much more judicious way to keep darker areas still dark but have the effects of light sources spread out further. We'll look into achieving this kind of getting the best of both worlds later, in the *Using global illumination* section.

ToneMap

This is something you can use as a quick solution for blending lights into darker areas, which will make everything look a bit more homogeneous. It comes with a few properties of its own:

- **Mode**: The default mode is **Linear**, and this is what you've been experiencing all along. We leave it to your taste, but we suggest you change it to **Filmic** or **ACES Fitted**. It'll remap the tones of the whole scene to the point that things will start to look more realistic.

- **Exposure**: Compared to **Linear** mode, the other modes may make your scene look really dark. Changing **Exposure** will brighten the scene while still applying the tone mapping.

- **White**: Digital cameras have a setting similar to this one. You designate a tone as white so that the other colors can be calculated according to this new baseline. Smaller values will blow out the whole scene because it'll start considering a lot more colors as white. Naturally, higher values will exclude more colors, and make the scene darker.

We won't mess with the **Exposure** and **White** values in our exercise, but this is what we have after choosing **ACES Fitted** for **ToneMap**:

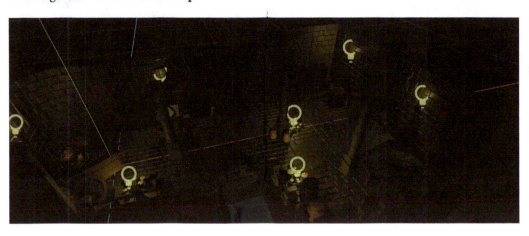

Figure 10.10 – Everything looks more pronounced thanks to tone mapping

Since we've touched on the concept of exposure, a quick word about enabling **Auto Exposure**. We won't use it in our work, but it is a helpful option for mitigating some of the problems you may face when the camera transitions between indoor and outdoor areas.

Screen Space Reflections (SSR)

When some objects have reflective qualities due to their material settings, such as **Metallic**, **Specular**, and **Roughness**, turning this environment setting on will create a more realistic effect.

To appreciate the impact of **SSR**, the level must have more light, so it may not look like much is changing when you turn it on. The body of the statue has a reflective material. Thus, if you zoom into that area, you should be able to see some reflection where the feet of the stag meet the pedestal.

Reflections will be more pronounced when there are more lights nearby. When we work on the player character's involvement in *Chapter 12, Interacting with the World through Camera and Character Controllers*, and Clara walks by the statue with a torch in her hand, you may notice the effect even better. Until then, we'll simply have this feature enabled.

Ambient Occlusion (SSAO)

This isn't the first time we have come across this term. We first got to know it in *Chapter 4, Adjusting Cameras and Lights*, when we wanted to emphasize the edges of the objects where they connected. Similarly, we'll turn this setting on in Godot too, but we have to tweak a few properties:

- **Light Affect**: You won't see the effect of **Ambient Occlusion** without the contribution of this property, so we are describing it first. It's for adjusting the role of light sources in the occlusion. We'll set it to 1.0.

  It seems as if we are using it as an on/off switch in our current situation. However, since it can be any value between 0.0 and 1.0, you can use it as a useful scale by controlling the value with scripts. This works in cases where you don't want to fully turn off the occlusions but gradually decrease them.

- **Radius**: When objects are close to each other, the contact points will look occluded. This setting is for adjusting the area that will be considered in the calculation for creating the correct amount of occlusion. We picked 0.4 as our value, but you can set it to any value, depending on your taste.

  Additionally, the **Intensity** property can be used with **Radius** to create more accurate occlusions. Also, with the help of a secondary set of radius and intensity, you can overlay more details.

As with most things in game development, adjusting the correct amount of Ambient Occlusion is often an artistic endeavor. With the suggested values, the result will be as follows:

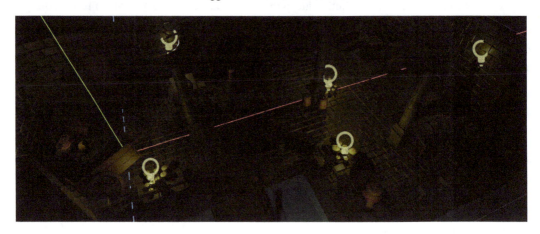

Figure 10.11 – The level after Ambient Occlusion has been turned on

The preceding screenshot may not be doing what we have achieved much justice. However, if you compare the previous two screenshots, you can see the occlusion in between the bricks, and also where the crates are making contact with the floor.

Glow

This feature is often referred to as the bloom effect in other applications. It's used to exaggerate the effect of colors, and especially light sources. While it has many properties, we'll only focus on a couple:

- **Bloom**: A value such as 0.2 will be enough to accentuate the effect of the sconces and the candles. In essence, while dark areas will stay relatively dark, lit areas will be glowing.

- **Blend Mode**: To increase the impact further, we suggest that you set this to **Additive**. It'll give the lights in the scene a real nice cozy effect since the light sources are open fires.

We won't touch the rest of the settings. The following screenshot shows the final state of the level:

Figure 10.12 – Our light sources glow in the dark

In the **Glow** settings, there is a particular section called **Levels**. You can expand that area and decide how far out the bloom and blur effect will emanate. It's useful when you want to adjust the detail of the bloom that's engulfing an object.

Adjustments

While applying different environmental effects, some of the features will be competing against each other. Even though we have more oomph for the lights, and more defined contours and shadows for the models, after a while, you may end up with a scene that looks a bit washed out. You will employ two properties of the **Adjustments** feature that will give your scene a decent touch:

- **Brightness**: Our main tool to remedy the washed-out look is increasing the contrast. However, turning up brightness alongside contrast works better. Feel free to adjust it the way you like it, but a value such as 1.1 or 1.2 might be enough.

- **Contrast**: This will tidy up the dull look and give the whole scene a more vibrant look. Using a value such as 1.1 will make things look better in tandem with more brightness.

We could go on forever while changing so many of these settings. Depending on your taste, you may prefer different effects. However, we are content with what we have so far.

Wrapping up

Our level's look has changed drastically since we first started laying out the floor and wall pieces. Ordinary-looking brick surfaces now have character, and the scene looks more ominous, thanks to lights, shadows, and finally, the environment settings. This can be seen here:

Figure 10.13 – The post-processing effects are all in place and working together

Depending on the atmosphere you want to create for your game, you can come up with a different combination of post-processing effects. Also, you can adjust their values programmatically during a game session to entice the player even more.

When one is too many

Post-processing effects are nice. You may feel like a kid in a candy store. However, keep in mind that some effects will enhance each other, and some will outdo each other. At the end of the day, you may end up having too many effects in play that are a burden on your computer. You can hear the cost of it when your GPU is vehemently trying to cool off.

Despite our efforts to improve the look of our level, there is room for improvement. While we have noticeably enhanced dark and bright areas, the scene is still missing another real-life quality that is often referred to as global illumination in the industry.

Using global illumination

If you've ever used a digital camera, you may already be familiar with the concept we are going to present in this section. Our brains, through expectation and familiarity with a similar environment, will blend in the light with darker areas, and fill in the missing parts. A camera, on the other hand, doesn't have prior knowledge of how places must look, and it can't process dark areas as well as our brains. In other words, the human brain approximates the missing parts and paints a more complete picture.

The rendering engine has worked like a camera so far. If you look at the level now, you'll see that the arched door is in the dark. It would be nice to have certain areas look more like what we would expect them to look like. If we increased the intensity of the light sources, it would cast the light farther away. However, we'd still end up with some areas darker than others. We need something that extends the effects of the existing light sources similar to the way our brains process light.

To that end, we'll introduce global illumination to achieve a more realistic look. Via this method, the area near the arched door will look like it's getting more light from nearby candles and sconces. If you haven't guessed it already, there is a node for this job. Let's add it to our scene:

1. Select the root node of the level.
2. Add a **GIProbe** node.
3. Adjust **Extents** in the **Inspector** panel so that **x** is 12, **y** is 5, and **z** is 15.
4. Turn its **Interior** setting on.
5. Position this probe in your level so that it engulfs everything like an envelope.

GIProbe will resemble a green wireframe cube initially. After you place it so that it wraps around the level, the Godot interface will look as follows:

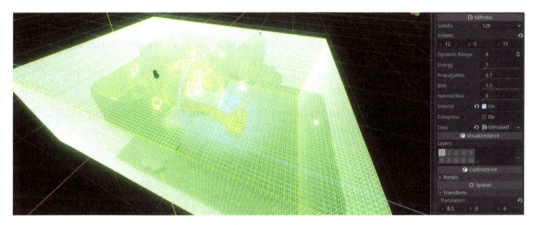

Figure 10.14 – GIProbe is in place but it's not functional yet

This node will probe the light sources in its volume. Then, it will interpolate this information to darker areas so that the light can be distributed more evenly, just as our eyes would expect. Although the probe is ready, we need to take care of two important things before we trigger the calculations.

Turning on Light Baking

We have already seen some of the import settings relevant to 3D models. For example, we saw that materials are imported automatically because it's the default setting in the **Import** panel. Also, using the **Animation** section in that panel, we were able to extract the actions from a model into the filesystem. All this was covered in *Chapter 7, Importing Blender Assets into Godot*.

We'll revisit the **Import** panel for a different need this time. We want some of the models to receive more light. So, by turning **Light Baking** on, some models will receive extra lighting information that's been sent by **GIProbe**. As the name suggests, this technique will bake some of the light in the scene into a model's material once. Then, it'll get updates as needed when the light conditions change.

So, we'll pick a list of models that look like they could benefit from light baking since they have large, uninterrupted surfaces:

- **Wall** (**Wall_Hole**)
- **Curve**
- **Floor_Standard** (**Floor_Standard_Curved_1** and **Floor_Standard_Curved_4**)
- **Column_Round**

Smaller objects such as props are usually not good candidates for light baking, but technically, you can turn the setting on for any model you import. For now, we'll select the wall model and enable light baking for it:

1. Select `Wall.glb` in **FileSystem**.
2. Bring up the **Import** panel and scroll down to find the **Light Baking** option (hint: this is the last option in the **Meshes** section).
3. Change its value to **Enable**.
4. Click the **Reimport** button.
5. Repeat this process for the other aforementioned models.

Generally speaking, we are enabling light baking for the architectural models in the scene. This is one part of the equation. Now that we have configured the models to accept light baking, we have to tell the renderer how much light should be baked into the materials for these models. We'll do that by adjusting the energy levels of the light sources we have used so far.

Adjusting Indirect Energy

The second most important thing in having proper global illumination is to adjust the energy levels of the light sources. Although this section's title indicates that we'll be adjusting indirect energy levels, it would also be useful to talk about what direct energy means.

In Blender, you changed the direct energy level for lights by adjusting their **Power** properties, which were measured in Watts. That meant you could have typed in real-life lightbulb values to get an accurate result. Godot's energy values for lights don't follow a unit system. So, it's more of an artistic value you can adjust based on your scene and liking.

While the **Energy** property, also known as direct energy, defines how intense the light will be, its **Indirect Energy** value is used to calculate the natural effect we described earlier in the opening lines of the *Using global illumination* section, where we made a comparison between human sight and cameras.

There is a simple way to observe this effect at home when it's sufficiently dark. You can light a candle and observe that there is going to be an adequately lit area near it. Then, the light will drop off gradually into the distance, but you'll still be able to notice some faraway objects. Their details won't be quite clear, but their most characteristic shapes will be apparent to the eye. It's possible to simulate this kind of effect with indirect energy using **GIProbe**.

For this effort, we have to adjust some of the **OmniLight** nodes we have used so far:

1. Open Sconce.tscn and select its **OmniLight** node.
2. Change its **Indirect Energy** to 2.5 under the **Light** section.
3. Change its **Range** to 8 under the **Omni** section.

This will increase the range of the light that's emanating from the sconces so that it'll reach farther. The energy level of **1.0** has already been used, hence we're only adjusting the indirect energy since we want it to contribute to global illumination.

Let's repeat this effort for the candles with different values:

1. Open Candles_1.tscn and Candles_2.tscn and select their **OmniLight** nodes.
2. Change **Indirect Energy** to 1.5 and **Range** to 3.

Compared to sconces, candles shouldn't emit that much light. So, it makes sense to have lower values. However, since there isn't one candle but a group of candles, the values aren't too far off. This is something you may have to balance in your work too: artistic concerns versus realism.

We've been settings things up for **GIProbe** to do its job. It seems like we have increased the overall light in the scene. We need it to be that way since some of this extra light will go toward calculating a better light distribution. All there is left to do is trigger **GIProbe**:

1. Select **GIProbe** in the **Scene** tree.
2. Click the **Bake GI Probe** button in the header just above the 3D view.

Godot Engine will calculate how light bounces off the surfaces of the models for which you have enabled light baking. Depending on the intensity, range, and indirect energy of the lights, the darker areas will receive more light. This will result in a more even distribution and give a more realistic look that meets our expectations. *Figure 10.15* shows the before and after of what global illumination does for the area near the arched door:

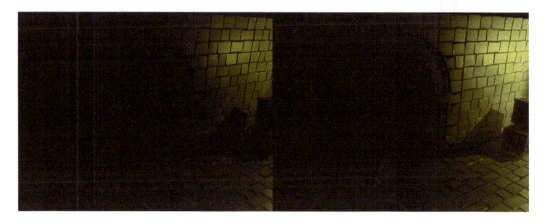

Figure 10.15 – The door has become more noticeable thanks to more evenly distributed light

Depending on the size and layout of your levels, you may need to place multiple **GIProbe** nodes. For example, if you were designing a dungeon with many rooms and hallways, it might be a better idea to consider each room and hallway as a unique **GIProbe** node since the distribution of lights will be achieved more accurately.

Also, when you have a level where an outdoor environment is connecting to an indoor environment, it's a good idea to create one **GIProbe** for each area and adjust the **Interior** settings accordingly. Using one major node that encompasses the whole level will do an injustice to either environment, so introduce as few and, sometimes, as many as necessary.

With that, we have improved the look of our level. Let's summarize the steps we have taken to get here.

Summary

The level we took over from the previous chapter looked complete, and yet uninteresting. To give it more life, we introduced a few instruments in this chapter.

First, we introduced two types of light nodes, **OmniLight** and **SpotLight**, to simulate candles, sconces, and the sun's effect in the cave. While accomplishing this, you also saw the reason why creating a scene for a model might be useful, as well as necessary, compared to instancing the models directly in the level. This effort was followed by adding a small script that can help you switch the lights if needed. We'll utilize this functionality later in this book.

Though lights were an obvious tool for improving the visuals, we also investigated shadows. They are resource-intensive, so you may want to turn them on for the lights that will have an important impact on your scenes.

To truly appreciate the effect of lights and shadows, we applied a bunch of environment settings. Although this helped the visuals a great deal, to elevate the realism to the next level, you've been introduced to global illumination. By carefully choosing which models should receive more indirect light and adjusting the setting of the lights in the scene, you've shed more light on certain areas, which resulted in a more accurate representation.

In the next chapter, we'll work on a different kind of visual system. It's a useful mechanism with which players can interact with the world: user interfaces.

Further reading

Out of all the topics we have presented in this chapter, global illumination is the most technical one. Simulating real-life light is a challenging task, and professionals out there are still actively working toward this goal. If you want to get a taste of it, here are a few links that should give you a better idea about what it involves:

- `https://ohiostate.pressbooks.pub/graphicshistory/chapter/19-5-global-illumination/`

- `https://www.scratchapixel.com/lessons/3d-basic-rendering/global-illumination-path-tracing`

- `https://developer.nvidia.com/gpugems/gpugems2/part-v-image-oriented-computing/chapter-38-high-quality-global-illumination`

On a more practical note, the official Godot documentation might be useful if you wish to learn more about what we have covered in this chapter:

- `https://docs.godotengine.org/en/3.4/tutorials/3d/lights_and_shadows.html`

- `https://docs.godotengine.org/en/3.4/tutorials/3d/environment_and_post_processing.html`

- `https://docs.godotengine.org/en/3.4/tutorials/3d/gi_probes.html`

11

Creating the User Interface

To start this chapter, let's begin by asking a simple question: what was the first multiplayer game you played?

If you are thinking of a PC or a console game, try thinking another way. Imagine a bunch of kids holding their arms out, pretending to shoot and take down the bad guys invading their neighborhood. Perhaps there was an evocative action movie the night before on TV. Now, these kids are bringing to life what they think is possible within the realm of physics, mixed with a bit of fantasy and what they remember from the movie. Some kids will even pretend they have been harmed along the way. Fallen comrades will be avenged in the end, and good will once again prevail against evil. Who's keeping the score here – that is, who has how many hit points?

How about the servers, internet speed, and likewise? Did the kids even need a **user interface** (UI) to play their game? No, because it was still easy for them to keep track of what was happening. But when the number of things people need to pay attention to gets beyond a certain point, it gets overwhelming. In other words, a UI is needed when using a system without one becomes impractical.

This is not unique to video games. In the real world, you use an ATM to access your bank accounts. The information and functions you need will be presented in a clear, concise manner; checking your accounts, sending e-transfers, and accessing the current interest rates are quick and easy to do all from one place, thanks to a well-designed UI.

In our game, despite what Clara expected, her uncle was not there but had left a note on the pier. We need a way for the player to access this information. Thus, in this chapter, we'll present a few of the UI components Godot has in its arsenal to convey this message. We'll start with a simple **Button** node, followed by a **Panel** component. In this panel, we will display some text via the **Label** component.

While you are adding more and more UI elements to the game, you'll also learn how to apply styles to these so that they look more like they belong to the game world. After all, the default ones have that default gray look, which might be better suited for prototyping.

Styling Godot nodes may feel tiresome after you do it more than a few times, especially if you are doing it for the same kind of button with different text. As a solution, we'll demonstrate how to take advantage of themes, which is a powerful tool that will help you in your styling efforts.

As usual, we'll be discussing a few relevant side topics that are pertinent to the creation of UIs. With that in mind, in this chapter, we will cover the following topics:

- Creating a simple button
- Wrapping in a panel
- Filling the panel with more control nodes
- Taking advantage of themes

By the end of this chapter, you'll have learned how to exploit UI nodes to help the player read the note that Clara's uncle had left for her.

Technical requirements

If you think you don't have enough artistic talent to create UIs, then rest assured for two reasons. First, we'll mainly focus on utilizing the UI components in Godot. Therefore, the graphic design aspect won't be our concern. Second, we are providing you with the necessary assets in the `Resources` folder in `Chapter 11` of this book's GitHub repository. Inside it, you'll find two folders: `Fonts` and `UI`. Simply merge these two folders into your Godot project folder.

This book's GitHub repository, `https://github.com/PacktPublishing/Game-Development-with-Blender-and-Godot`, contains all the assets you need. Lastly, you can either continue your work from the previous chapter or utilize the `Finish` folder from `Chapter 10`.

Creating a simple button

A UI is a collection of components you lay out in a coherent manner around the core visuals of your game. The most essential UI component to start with may have been a **Label** node if we wanted it to be similar to printing "Hello, world!" when we are learning a new programming language. However, we'll start with a **Button** node since the former case is so trivial, and we can also learn how to style a **Button** during this effort.

Before we start throwing around a bunch of UI nodes willy-nilly, we should first mention the right kind of structure to hold our UI nodes. We can use **CanvasLayer** similar to using a **Spatial** node to nest other nodes such as **MeshInstance**, **AnimationPlayer**, and others.

We've already been creating scenes mainly to display 3D models. Let's follow similar steps for the sake of creating the UI:

1. Create a blank scene and save it as `UI.tscn` in the `Scenes` folder.
2. Choose **CanvasLayer** for its root node and rename it `UI`.
3. Attach a **Button** node to the root and rename it `Close`.
4. Type `Close` for its **Text** value in the **Inspector** panel.

There's nothing fancy going on so far, but we now have a button aligned, by default, to the top left of the viewport. The width of this button also expanded when you were typing the text it displays.

> Control versus CanvasLayer
>
> We mentioned that a **Spatial** node would be the root node for 3D nodes. So, for the sake of keeping things familiar, we could have used a **Control** node to hold the **Button** node. Rest assured, you could still inject a **Control** node inside a **CanvasLayer**. The real reason we used a **CanvasLayer** as the root is for its **Layer** property in the **Inspector** panel. By changing the value of this, you can change the draw order, which means you can decide which **CanvasLayer** will render first. This is a useful mechanism when you have multiple UI structures that need to be layered on top of each other in a precise order.

The button we have just added looks boring. It doesn't quite fit the world we are creating. Now, let's use a custom graphic asset to style our button:

1. Expand the **Styles** subsection in the **Theme Overrides** section of the **Inspector** panel.

2. Using the dropdown for the **Normal** property, select the **New StyleBoxTexture** option.

3. Click the **StyleBoxTexture** title as it will populate the **Inspector** panel with its properties.

4. Drag `button_normal.png` from UI into the **FileSystem** panel and drop it in the **Texture** property.

5. Expand the **Margin** section and type 8 for all the margin values.

6. Press *F6* to launch the `UI.tscn` scene and try to interact with the button.

You have taken quite a few steps to style a simple button, so let's break down what's happened.

In *step 1*, you told Godot that you wanted to override the default theme, which was giving that gray look to the button. Without user interaction, the button will be in its normal state; so, that's what you intend to change in *step 2*. We'll discover how to change the other states very soon.

Step 3 was about defining the properties of this **Normal** state. For this, you used an aptly named texture file in *step 4*. Then, in *step 5*, you adjusted the margin values so that the texture permitted the text to have enough room without snapping to the edges. For example, try to change the text of the **Close** button to `Lorem ipsum dolor sit amet`. Notice how the button is getting wider without looking stretched and keeping the rounded corners intact. This needs a proper explanation.

Setting margins involves doing more than just accommodating text. Carefully selected values will make sure the texture will enlarge or shrink as needed without losing some of its qualities, such as rounded corners. When the asset has rounded corners, if the texture is stretched, you will end up with a distorted look. The practice of conserving the core features of a texture and allowing it to be resized properly without distortion is called 9-slice scaling. You can learn more about it here: `https://en.wikipedia.org/wiki/9-slice_scaling`.

When you launched the UI.tscn scene in *step 6*, the button must have shown its normal state as a brown texture. If you move your mouse over it, you'll see that the button will show the default look again because you haven't set the hover state yet. This can be seen in the following screenshot:

Figure 11.1 – The button only has its normal state styled

Similar to the way you assigned a texture to the normal state of the button, you can do so for the other states. Let's do this for the hover state:

1. Select the **Close** button in the **Scene** tree.

2. Assign a **New StyleBoxTexture** to the **Hover** state in the **Styles** subsection under **Theme Overrides** and click this **StyleBoxTexture** to set its properties.

3. Drag button_hover.png from the UI folder and set the margins to 8.

4. Press *F6* and move your mouse over the button.

We'll repeat this effort for the pressed and disabled states as well. We won't use disabled buttons in our game, but why not be thorough? Also, in most scenarios, you can repurpose the pressed state for the focus state. The different results are shown in the following screenshot:

Figure 11.2 – The normal, hover, pressed, and disabled states of a button with a custom texture

Before we move on to introducing more UI nodes, we suggest that you change the text of that button back to Close since we'll use this button to close a panel that will simulate a note from Clara's uncle. Speaking of which, it's time to learn what was written in that note.

Wrapping in a panel

So far, we have created a button and styled it. However, it would be nice if it served some purpose, especially since we gave it a meaningful label. We'll write some code so that this button can close a panel near the end of the *Filling the panel with more control nodes* section. Before we get to that point, though, we need the panel.

As we are introducing more UI nodes, let's remember why we are doing this within the game's context. Clara's uncle had left a note. We'll simulate that note with a combination of UI nodes in Godot so that it looks as follows:

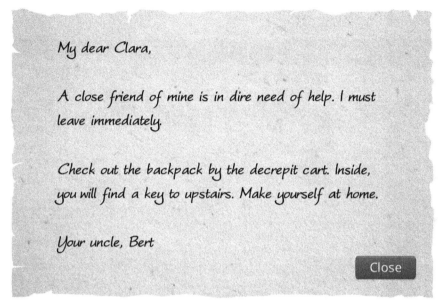

Figure 11.3 – Clara's note

We've already taken care of the button, but it is currently sitting in the middle of nowhere. We'll wrap it in a **Panel** node in this section after we give a short disclaimer.

A **Panel** node is just another **Control** node in Godot that usually holds other components. There is a similarly named node, **PanelContainer**, which might be confusing for beginners. The **Panel** node derives from the **Control** class, whereas the **PanelContainer** node inherits from the **Container** class. Also, the **Container** class inherits from the **Control** class. This kind of technical detail might be important when you are doing more advanced work. We won't, so either one would work fine for our intents and purposes in this book. Therefore, we'll stick with the **Panel** node.

At this point, we are ready to add a **Panel** node and style it:

1. Add a **Panel** node under the root **UI** node in the **Scene** tree.

2. Expand the **Rect** section in the **Inspector** panel.

3. For the **Min Size** property, set the following values:

   I.   Type 600 for **X**.

   II.   Type 400 for **Y**.

4. Assign a **New StyleBoxTexture** to the **Panel** property in the **Styles** subsection under **Theme Overrides**.

5. Drag the **Close** button over the **Panel** node in the **Scene** panel so that the **Close** button is nested.

At this point, you should have the following output:

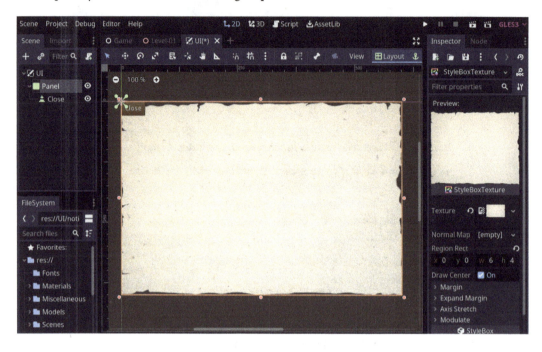

Figure 11.4 – The paper texture has been simulated with the help of a Panel node

We are getting closer and closer to the desired design we imagined for the note. The button in the panel is still aligned to the top left. You can drag it to a position that makes sense, but it might be easier to decide on that if you have some text within the panel. That's what we'll take care of next.

Filling the panel with more control nodes

The uncle's note is slowly taking shape. We'll introduce a **Label** node in this section for the text portion. Additionally, we'll have to figure out how to position all these elements so that the note resembles the layout we'd like to have. Lastly, we'll discuss a few complementary **Control** nodes you may want to use in some other scenarios.

After all, we will still employ the most basic UI node: **Label**. If we had used it at the beginning, it would have looked unimpressive with its default style and color. Since we now have a proper texture over which this **Label** node can go, things will look more interesting. Follow these steps to do this:

1. Select the **Panel** node in the **Scene** panel.

2. Add a **Label** node and turn its **Autowrap** property on in the **Inspector** panel.

3. Set its **Text** to the following:

```
My dear Clara,


A close friend of mine is in dire need of help. I must leave
immediately.


Check out the backpack by the decrepit cart. Inside, you will
find a key to upstairs. Make yourself at home.


Your uncle, Bert
```

Our last effort will result in an awkwardly tall text block. To remedy this, we could manually give some width and height to the **Label** node we have just inserted. While we are doing that, we could also change its position to make it look centered and have some margins off each edge. However, we can do something smarter: we can wrap this **Label** inside a **MarginContainer** that will set margins and automatically resize the text for us.

Adding a MarginContainer

At this point, adding new nodes to the **Scene** panel must be a common task for you. Nevertheless, there are times, such as now, when deciding where to add a new node and what to nest in it might not be obvious. The question is, where can we add **MarginContainer**? Outside the **Panel** node or inside?

A **MarginContainer** is a specialized container that's responsible for introducing margins so that its children look like they have padding. If we wrap the **Panel** node inside a **MarginContainer**, since the **Panel** node is holding the text, the whole structure, including the button, will be padded. That's not good since we would like the text to have some space between its edges and the borders of the texture that constitutes the **Panel** node. Thus, this is what you need to do to only pad the text:

1. Add a **MarginContainer** node inside the **Panel** node and nest **Label** inside this **MarginContainer** node.

2. Set the following values in the **Inspector** panel for **MarginContainer**:

   I. In the **Anchor** section, set both **Left** and **Top** to 0 and both **Right** and **Bottom** to1.

   II. In the **Margin** section, set all its properties to 0.

   III. In the **Constants** subsection under the **Theme Override** section, set both **Margin Right** and **Margin Left** to 60.

We touched on a lot of terms in the preceding operation. The first two sets of actions, where we alter the values of anchor and margin, are not specific to a **MarginContainer**. They exist for every type of **Control** node. You can also see this fact as these properties were listed under the **Control** header in the **Inspector** panel.

The anchor and margin values we chose are such special values that we could have used a shortcut to achieve the same result. It would be selecting the **Full Rect** option in the expanded menu after you click the **Layout** button in the header section of the 3D viewport. This **Layout** button is visible in the following screenshot, just above the top-right corner of the paper texture.

We'll use another option under that menu when we adjust the location of the **Close** button later. For now, compare your work to what you can see in the following screenshot:

Figure 11.5 – The text now has padding, although it's hard to read

What was essential in the properties of that **MarginContainer** was adjusting its content margin values in the **Constants** subsection. That gave the text some room and positioned it correctly over the paper texture.

It's a bit difficult to read the text, though. So, let's see how we can make it legible and, even better, make it look like *Figure 11.3*.

Styling the Label node

Although **MarginContainer** is now occupying as much space as the **Panel** node, and it's providing margins to the text it's holding, the text itself is hardly legible since it's small and white over a lightly colored surface. Also, the font choice is wrong because it's using the default font provided by Godot Engine. We'll learn how we can fix all these issues in this section.

Let's start by selecting the **Label** node in the **Scene** panel so that we can make some changes under **Theme Overrides**:

1. Turn on the **Font Color** option in the **Colors** subsection. The color can be left black.

2. Choose the **New DynamicFont** option for the **Font** property in the **Fonts** subsection and expand this option's properties right away by clicking its title. We need to edit the subsections:

    I.   Drag `Kefario.otf` from **FileSystem** to the **Font Data** property in the **Font** subsection.

    II. Change **Size** to 28 in the **Settings** subsection.

We'll discuss what's happened shortly, but here is what we have done so far:

Figure 11.6 – The Label node now looks more like handwritten text

The default black color for the text seems to be fine, but you could pick a different color if you wish. A much more drastic change happened when we introduced a font type. We did this in two steps. First, we picked a **DynamicFont** type, which is slower than the other option, **BitmapFont**, but it lets you change the properties of the font at runtime. However, this is not enough to render a font since it works like a wrapper. So, you need to assign the font you would like to render. That's what we did when we assigned a font file to the **FontData** property.

There is an important caveat we think you should be aware of with fonts since they are made of individual elements called glyphs. You can think of them as the letters in an alphabet. Not every font supports the full spectrum of an alphabet. For example, in the note UI that we designed, if you replace the text you will with its shortened form, you'll, the apostrophe won't render because it doesn't exist as a glyph in the font. Usually, paid fonts come with a bigger set of glyphs. Otherwise, keep searching for free options with a more complete set.

> **Pixels versus points**
>
> When we chose 28 as the font size, that number was measured in pixels. In some graphics or text editors, you'll often find fonts measured in points. This is something you have to be cautious about because if you transfer the numbers verbatim to Godot, your font will be rendered quite differently. So, mind your units.

In the real world, a note from Clara's uncle would only contain the text portion. Thus, it would be absurd to expect a close button on top of an actual piece of paper. However, this is a game, and we've already discussed how UIs mix reality with functionality. To complete the UI for the note, it's time we positioned that button.

Positioning the Close button

We used a nice trick to position the text concerning the piece of paper it's on. Can we replicate this for the **Close** button? Since a button can't be considered a wide structure, we can't stick it inside **MarginContainer**. However, we can still position it relative to the **Panel** node.

In the *Adding a MarginContainer* section, we used a longer method to adjust the dimensions of that component. We also mentioned that we would use a shortcut. This is how you can use it after selecting the **Close** button in the **Scene** panel:

1.   Expand the **Layout** menu and select the **Bottom Right** option.

2.   Hold down *Shift* and press the left and up arrow keys on your keyboard four times for each.

This will position the **Close** button at the bottom right corner, then pull it up and move it left just enough that it stays there. We mean it when we claim that it'll be staying there. For example, select the **Panel** node, then try to resize it using the handles in the viewport. Does the button stay nicely tucked in that bottom right corner? Good! How about the **Label** node? Does the text flow to occupy the extra space? Neat!

Our efforts to develop what you saw in *Figure 11.3* are coming to fruition, as shown here:

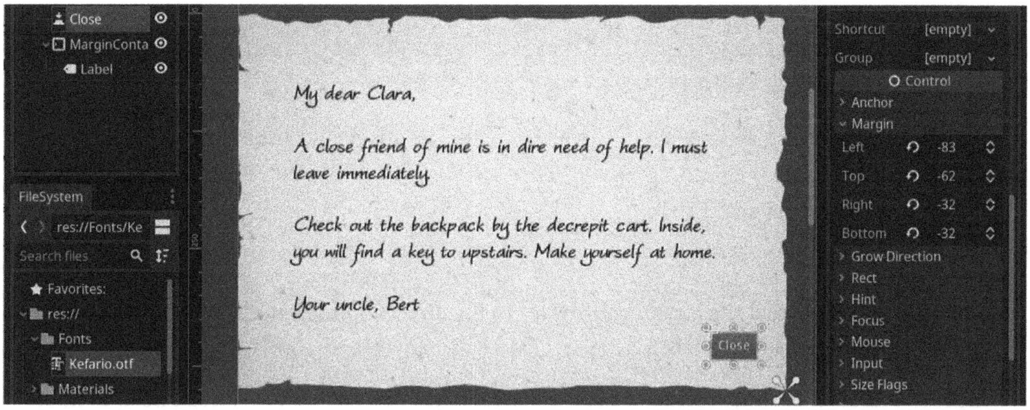

Figure 11.7 – Everything in the UI is positioned carefully

If you want to test your scene, go ahead and press *F6*. Depending on your setup, you may notice that the **Close** button will not be functional since it's behind **MarginContainer**. So, try to resort the nodes in the **Scene** panel by dragging the nodes up and down. When you have the **Close** button after **MarginContainer**, everything should be functional.

Speaking of functionality, we haven't wired anything up for the **Close** button. Ideally, that button should turn the visibility of the **Panel** off so that the note looks as if it's been closed. Let's do that next.

Adding the close functionality

There are multiple ways we can attack this problem. We are going to show you one for brevity's sake so that you can see what's involved. You may have to apply similar principles differently in your future projects.

For example, the way we are treating the UI.tscn scene so far is to have one big **Panel** node as a direct child. Your games may need a lot more UIs with elements permanently visible on the screen, more notes to open and close, inventory screens with expanding parts to reveal more details, and likewise. There are many possibilities, which is why there are different types of architectures you can construct. There will always be a tradeoff between these different options, so we suggest you experiment with the benefits of different UI structures if you have some spare time.

Without further ado, our suggestion for implementing the closing functionality is to add a small script to the **Close** button. Select it and do the following:

1. Attach a script to the **Close** button and save it as ButtonClose.gd in the Scripts folder.

2.   Make this script file look as follows:

```
extends Button

func _ready():
    connect("pressed", self, "on_pressed")

func on_pressed():
    get_parent().visible = false
```

This architecture assumes that the button will always be the direct child of a node, so once it's pressed, it will make its parent invisible. Ouch!

The benefit of this kind of simple structure is the convenience that the button doesn't need to know the node structure it's in. There is also a more conventional way of attaching the *pressed* behavior by using the **Node** panel and binding a signal. Either way is fine.

Constructing and improving UI elements may easily turn into a project by itself. You might be tempted to create that perfect setup for all future possible scenarios but keep in mind that overoptimization is a thing. Later, you may realize that you didn't need all that preparation in the first place. We'll talk about a similar situation next, where the note might be longer.

Wrapping up

We now have a fully functional UI for displaying the note from Clara's uncle, Bert. What if Bert had more to say? For example, let's say the message had an extra line after his name, as shown here:

```
Your uncle, Bert

P.S. I think I might have left my pet snake unattended. It might be
wandering around, so be careful!
```

If you were to add this extra text to the end of the **Label** node, the text would get uncomfortably close to the top and bottom of the paper texture. Similarly, imagine that this text block needed to be even longer, which is the case in some types of games where exposition is important. For instance, it is very common when displaying the details of a quest or an item in role-playing games.

Currently, we can make do by adjusting the font size of the text or making the margins narrower to allow more room for the new text. However, in more extreme situations, it might be better to use a **ScrollContainer** node. Just like you wrapped the **Label** node inside **MarginContainer**, you can wrap a **ScrollContainer** around the **Label** node, and tweak a few things to have a scrollable text block.

Coming up with the correct level of *nestedness* and deciding on the type and order of UI components is sometimes an effort of trial and error. Consequently, there aren't any set formulas. Therefore, you may find yourself practicing and seeing what works best in your use case.

That being said, generalizing your efforts to maintain a consistent look and feel across your many UI nodes might be helpful. We'll tackle themes next to accomplish this.

Taking advantage of themes

Using or, more specifically, creating themes in your projects is smart on many accounts. First, we'll discuss their usefulness, show you a few visual examples, and then create one for practicing. Let's start with the reasons why you should use themes.

Firstly, using themes will save you from manually applying overrides to the components the way you've done so far. It's still possible to keep adding manual touches here and there, but what would happen if you wanted to change a button's artistic direction completely? This would trigger a chain reaction to change the look of other components too. So, you'd have to restart the manual editing. Furthermore, the ultimate worst-case scenario would be to revert your changes because, you know, we are human and we kind of tend to stick with our first choices more often than not.

Secondly, you could have multiple themes at the ready in your game. Although a button is still just a button, you could assign it one theme out of many. This will make that button look like it belongs to a family of components. Thus, your UI elements will have a consistent style.

Lastly, changing themes at runtime is a possibility. This means that if, in your game or the application you are building with Godot, you would like to swap themes for special occasions such as Christmas, this is entirely possible. Also, more and more desktop applications are being built with Godot. These could also benefit from theme swapping to offer their user the best choice. Godot Engine itself allows you to change themes. You can access the existing themes by opening **Editor Settings** and trying out a few themes. For example, try out the **Solarized (Light)** theme. Are you getting Unity vibes?

Changing a theme is not always about changing the colors of buttons or font sizes. For example, `https://365psd.com/psd/ui-kit-54589` and `https://365psd.com/day/3-180` are two UI kits we picked to show how drastically different your Godot UI nodes could look. *Figure 11.8* presents these two UI kits side by side:

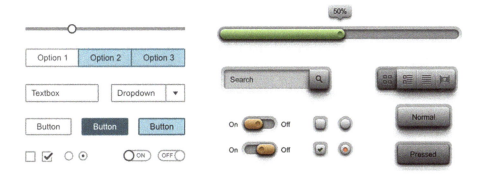

Figure 11.8 – Two distinct UI kits that are good candidates for themes

Since we have already seen how to change the look and feel of three types of nodes, **Button**, **Panel**, and **Label**, we'll focus on other types of **Control** nodes. We'll accomplish this in the context of creating a new theme.

Creating a new theme

Since game development is an iterative process, planning every single thing ahead of time may not always be possible, and even be fruitless. That's why it's typical if you start by changing the UI nodes manually. Still, starting with a new theme and changing the properties of this theme may also be a good idea. Why? Because if your experiments for individually modifying the components yield a successful result, you won't have to repeat what you have done in the theme. By creating a theme at the beginning, you're building up as you go.

Also, creating a theme is like creating any other type of resource in Godot. We can do this by following a few simple steps:

1. Right-click the UI folder in the **FileSystem** panel, choose **New Folder**, and type Themes as its name.

2. Right-click Themes in **FileSystem** and select the **New Resource** option.

3. Choose **Theme** as the resource type and save it as Dark.tres.

This will create a **Theme** resource in your project. It should also enable a new panel in the bottom area that will show the preview of this new theme. As you make changes to your theme, updates can be previewed in this area since it might be faster to monitor your progress this way rather than adding and removing test components to/from your scene.

If the preview area looks small, it's possible to enlarge it by clicking an icon next to Godot's version number. This icon will look like two upward-facing arrows with a horizontal line above them. Press that and the theme preview will occupy the viewport. In the end, your editor will look as follows:

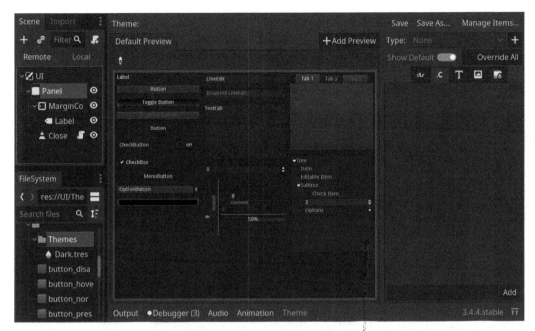

Figure 11.9 – The theme preview has been expanded

By the way, the preview area is not static. You can interact with those UI components. It's like a Godot scene running inside Godot. Now, we will modify the theme for the **CheckButton**, **CheckBox**, and **VSlider** components. We'll also show a special state of the **CheckBox** node, also known as a radio button, in web development. However, our first candidate is **CheckButton**.

Styling a CheckButton

The graphics assets we'll be using to construct the new theme is the *Dark UI Kit*, which you can find at https://365psd.com/psd/dark-ui-kit-psd-54778. We've already exported the necessary parts into the UI folder for you.

The theme we created is still the default theme, so it still shows the default components. We'll have to pick the one we would like to change. This is how we do it:

1. Press the button with the plus (+) icon in it. This is in between the **Manage Items** and **Override All** buttons in the top-right corner of the **Theme** preview area.

2. Select **CheckButton** in the upcoming pop-up menu. By doing this, you will see a list of this component's relevant properties separated by tabs on the right-hand side of the theme preview.

3. Switch to the fourth tab, which looks like a polaroid icon with a mountain in it. Press the plus (+) icons for the **off** and **on** properties.

4. From the **FileSystem** panel, drag `dark-ui-checkbutton-off.png` to the *off* slot and, similarly, drag `dark-ui-checkbutton-on.png` to the *on* slot.

5. Interact with **CheckButton** in the theme's preview.

This will effectively change the look of **CheckButton**. Your **Theme** panel will look as follows:

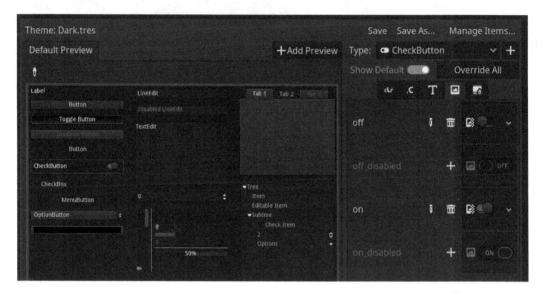

Figure 11.10 – We have changed the look of the CheckButton component with custom assets

CheckButton is a simple component with two main states: on and off. We were not interested in altering the disabled versions of its two states, simply because the UI kit does not have the assets for that permutation. If you think you'll never have this component in a disabled state, then you don't have to create and assign art either.

Let's attack a different component this time. Although its name is similar, and it comes with states similar to **CheckButton**, a somewhat disguised property of this node makes it function as two distinct components. Enter **CheckBox**.

Changing a CheckBox and discovering radio buttons

This is going to be a similar effort, but we'll utilize more assets and fill out more properties. Let's keep the momentum going and add a new item to the theme:

1. Using the plus (+) icon button again, choose **CheckBox** from the upcoming item list.

2. The fourth tab may still be active. If not, switch to it and do the following:

    I.   Assign `dark-ui-checkbox-off.png` to the **unchecked** property.

II.   Assign `dark-ui-checkbox-on.png` to the **checked** property.

III.   Assign `dark-ui-radio-off.png` to the **radio_unchecked** property.

IV.   Assign `dark-ui-radio-on.png` to the **radio_checked** property.

When you prepare your assets, pick filenames that are close enough to the state the assets will be assigned to. So, associating these files between the **FileSystem** and **Theme** panels would feel easy. After making these changes, this is what we have:

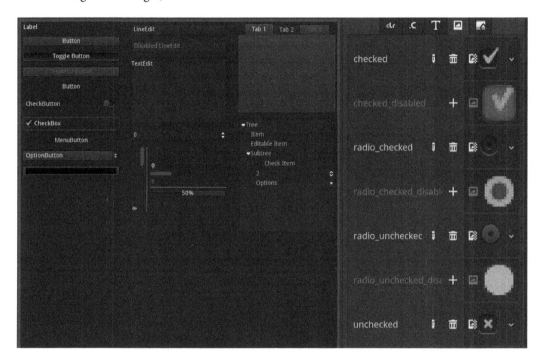

Figure 11.11 – CheckBox is the latest item we have customized for our Dark theme

The preview area has the **CheckBox** component for you to test, but no radio button. There is no **RadioButton** component in Godot. Despite adding the assets for it, we can't simulate it yet. Nevertheless, we can tweak a **CheckBox** component so that it acts like a radio button.

Since we need to physically place a **CheckBox** component in the scene, you can toggle off the button that maximized the **Theme** panel. Alternatively, you can press *Shift + F12*, and follow these steps to add a few components to the `UI.tscn` scene:

1.   Turn the visibility off for the **Panel** node by clicking its eye icon in the **Scene** panel.

2. Select the root node, then add an **HBoxContainer** node. Select this new node right away so that you can do the following:

    I.   Add a **VBoxContainer**, **VSeparator**, and another **VBoxContainer** to it.

    II. Add two **CheckBox** nodes inside these two **VBoxContainer** nodes.

    III. For the first two **CheckBox** nodes, change their text properties in the **Inspector** panel to `Multiple Choice 1` and `Multiple Choice 2`, respectively.

    IV. For the last two **CheckBox** nodes, change their text properties in the **Inspector** panel to `Single Choice 1` and `Single Choice 2`, respectively.

We're not done yet, but the following screenshot shows what's happened so far:

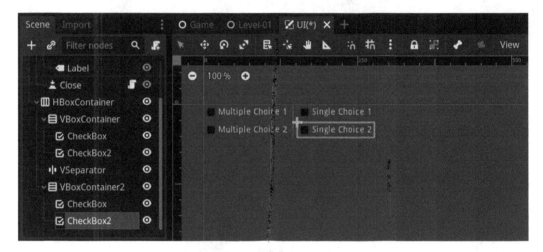

Figure 11.12 – Four checkboxes organized in a questionnaire fashion

We are a few steps closer to turning two of those checkboxes into radio buttons – specifically, the last two since we gave them some text that mentions a single choice. Thus, while you have **CheckBox2** in the **VboxContainer2** node selected, do the following:

1. Assign a **New ButtonGroup** to its **Group** property in the **Inspector** panel.

2. Click the down arrow in that **Group** slot to expand a dropdown menu and select **Copy**.

3. Select the **Checkbox** node in **VBoxContainer2** and choose the **Paste** option by expanding its **Group** options. This will link the two checkboxes because they will be sharing the same button group.

You should notice a drastic change between the two sets of checkboxes. Whereas the first two still look like checkboxes, the last two have circular icons next to them, as shown in the following screenshot:

Figure 11.13 – Two checkboxes have been converted into radio buttons

By sharing the same button group, checkboxes turn into radio buttons. In this exercise, it was sufficient to create and assign a generic **ButtonGroup** object. However, if you want to have a group of radio buttons in one area of your application, then another collection somewhere else that governs a different set of radio buttons, you may have to create named **ButtonGroup** objects and assign them accordingly.

We won't cover that kind of scenario since we seem to be missing something more important that we have wanted for a while. Neither the checkboxes nor the radio buttons we worked so hard for are reflecting the artistic direction we defined in our theme. Let's see how we can utilize our theme.

Attaching a theme

Previously, we mentioned that using themes would help you style components faster. It's true, but we haven't tested this claim yet. Since we've already prepared the styles for the checkboxes and radio buttons, all there is left to do is assign the theme to these components:

1. Select the **HBoxContainer** node in the **Scene** panel and expand the **Theme** section in the **Inspector** panel.
2. Drag Dark.tres from **FileSystem** to fill the empty **Theme** slot.

There you have it! We didn't even have to select each component and assign the themes one by one. A higher-level structure such as **HBoxContainer** was enough to assign the theme to so that its children could use the relevant parts.

Do you see the real potential here? Assigning a theme to a root element will be enough most of the time. That being said, since each component can be assigned its own theme, but it doesn't have to, you can have all sorts of permutations. In its simplest form, assigning a theme to a root node will be enough in most scenarios.

So far, we've been styling relatively simple UI nodes, such as **CheckButton** and **CheckBox**. Maybe we could tackle another node that has a few moving parts, such as a **VSlider**.

Altering a vertical slider component

A vertical slider component, **VSlider**, is useful when you want to give your players an easy way to adjust the ratio or quantity of things, such as tradeable items during a game session, music volume, or the brightness level in the game's settings. Likewise, you can use an **HSlider** node, which is the horizontal version, but both accomplish similar tasks.

Since we only have the graphic assets for a **VSlider**, we'll only cover this styling. If you desire, it's possible to convert the existing assets that are compatible with an **HSlider**. You'll have to rotate each part 90 degrees and save them accordingly. To do so, you must follow these steps:

1. Add **VSeparator** and **VSlider** nodes to **HBoxContainer** in the **Scene** panel.

2. Using the **Inspector** panel, type 75 for the **Value** property for **VSlider**.

3. Double-click `Dark.tres` in **FileSystem** to bring up its details. Add **VSlider** as a new type using the good old button with the plus (+) icon.

4. Activate the fourth tab in this new type's custom properties and assign `dark-ui-vslider-grabber.png` to both **grabber** and **grabber_highlight**.

5. Switch to the fifth tab, which looks like a square rainbow.

6. Attach a **New StyleBoxTexture** to the **grabber_area** property. Click the slot to see its details and do the following:

    I.   Assign `dark-ui-vslider-grabber-area.png` to the **Texture** property.

    II. Expand the **Margin** section and type 6 for the **Bottom** property.

7. Bring up the theme preview again by double-clicking `Dark.tres` or switching to the **Theme** panel at the bottom.

8. Instead of repeating the same effort for the **grabber_area_highlight** property, click the plus (+) button near its slot, then grab and drop the **grabber_area** property's style onto the **grabber_area_highlight** slot. Alternatively, you can copy the slot from **grabber_area** and paste it into **grabber_area_highlight** using the dropdown menus.

9. Attach a **New StyleBoxTexture** to the **slider** property. Click the slot to view its details and do the following:

    I.   Assign `dark-ui-vslider-slider.png` to the **Texture** property.

    II. Expand the **Margin** section and type 6 for the **Bottom** and **Top** properties.

    III. Make the **Expand Margin** section visible and type 1 for all its properties.

10. Press *F6* and admire your hard work.

We took many steps here, but there were only one or two new things. First, we repurposed one of the styles to be used for a different property by dragging and dropping it. This is a shortcut method instead of copying and pasting between slots. It's useful when both slots are near each other. If you are copying elements where the slots are on different panels, then you still have to resort to the copy and paste method in dropdown menus.

Secondly, we adjusted a different type of margin, **Expand Margin**. The slider has two separate parts that constitute its track where the scrolling occurs, so we had to adjust this special margin so that it fits the blue part inside the outer part. Take a look at the following screenshot; you will see that there is a blue filler under the grabber inside the track of **VSlider**:

Figure 11.14 – It took a few more steps but the VSlider component has been thematized

It's easier to see the effect live than reading it. So, when you launch the UI.tscn scene, try to interact with the grabber and see how the component fills its track with blue, depending on the position of the grabber.

Wrapping up

This concludes our work in setting up a theme. Although we have styled only a handful of nodes, feel free to practice with the rest of the same UI kit or pick another one from the website to try it on other **Control** nodes.

All in all, working with themes or individually styling components entails two things. Primarily, you can either assign textures directly to some of the properties or indirectly into the appropriate slot by creating a **StyleBoxTexture**. Secondly, there are some numerical properties you can tweak. We haven't covered this latter case. For example, you can adjust the line height of components that deal with text rendering. These cases are easy to comprehend and test. So, we opted to show you more head-scratching cases.

Hopefully, by practicing what we have shown so far and discovering more on your own, you will be able to apply beautiful graphic designs to your game.

Summary

We started this chapter by debating what UIs are. We did this via a brief philosophical and theoretical explanation.

Assuming your games will require UIs, we investigated a practical use case such as constructing a note left by Clara's uncle. This work necessitated us to work with multiple **Control** nodes – that is, the **Button**, **Panel**, and **Label** nodes.

During this effort, not only did we employ the components we needed, but we also styled them to match a specific artistic style.

For the sake of not repeating ourselves and taking the styling to the next level, we presented how using themes might be a time saver. To that end, we showed you how to utilize UI kits you could find online by assigning these kits' individually exported graphics assets to the properties of **Control** nodes.

UIs are, in a way, a tool for the player to interact with the game. That being said, in the next chapter, we'll discover a more direct way to interact with the game world without the help of UIs.

Further reading

In the introduction, we talked about when a UI is necessary. However, there are situations when the best interface is no interface at all. There is an app – sorry, a book – for that by *Golden Krishna*: *The Best Interface Is No Interface: The simple path to brilliant technology*. He talks about how introducing more steps and elements disguised as a UI is nothing but interference.

We've already discussed the possibility of having games without a UI, but we'll rest that argument for now. It might be better for you to be exposed to as much information and examples as possible at this point. So, the following are a few technical and practical resources:

- `https://www.toptal.com/designers/gui/game-ui`
- `https://webdesign.tutsplus.com/articles/figma-ui-kits-for-designers--cms-35706`
- `https://ilikeinterfaces.com/`
- `https://www.gameuidatabase.com/`

This chapter also showed you how to assign fonts to components. There are a lot of freely available fonts out there but be careful and read their licenses. They might be downloadable but some of them can't be used in commercial work. The same kind of warning goes for anything else too, especially graphics assets.

12

Interacting with the World through Camera and Character Controllers

You have been preparing little bits and pieces for the game world, especially in the last two chapters. In *Chapter 10*, *Making Things Look Better with Lights and Shadows*, you added **Light** objects to sconces and candles. You even placed a script to adjust these objects' lit state. Then, in *Chapter 11*, *Creating the User Interface*, you built a new scene by introducing **Control** nodes. This effort was for simulating a note from Clara's uncle, Bert.

Although we've been taking steps to make things more sophisticated, pretty much everything feels static. In this chapter, we'll show you a collection of practices that will build a connection between game objects and the player. This will make the project look live and feel more like a game.

The first thing we'll look at is the **Camera** node and its settings. Godot's viewport has been letting you see different scenes via a temporary construct so that you could work with the software. Such a transitory concept won't be enough, so we'll work with our camera system.

Next, we'll focus on building a connection between some of the game objects in the world and the player. This involves detecting mouse events on a 2D surface and projecting these events into a 3D space. There might be different interactions such as hovering, clicking, pressing, and likewise, so we'll look into ways to detect the action we want. For example, we will click a parchment left on the pier to bring up the note we worked on in the previous chapter.

Similarly, if the click happens to be on one of the areas where we would want to move Clara, we need a system that can do the pathfinding for us. To that end, we'll investigate new Godot nodes, **Navigation** and **NavigationMeshInstance**.

Lastly, why not add a bit of animation? After we discover how to move a game object between two points in the world, we could instruct this object to trigger the appropriate animation cycle. In our case, Clara will switch between her idle state to her walking state. As a result, we'll revisit some of the notions we got to know in the *Importing animations* section of *Chapter 7, Importing Blender Assets into Godot*.

As you can see, we are going to utilize a lot of the topics we have already visited, yet there is still a lot of new stuff to discover and learn. If we could enumerate it, it would look like this:

- Understanding the camera system

- Detecting user input

- Moving the player around

- Triggering animations

By the end of this chapter, you'll have a much better understanding of camera settings in general, and you'll be able to detect your player's intentions and relate them to actions in the game. Thanks to an easy method of pathfinding, you'll move Clara around the level to a location you want, and—finally—trigger the appropriate action to simulate her walking.

Technical requirements

We'll continue where we left off in the previous chapter. This means you can keep working on your existing copy. Alternatively, you can start with the `Finish` folder of `Chapter 12` in this book's GitHub repo: `https://github.com/PacktPublishing/Game-Development-with-Blender-and-Godot`.

We have several new assets that are necessary to do the work in this chapter. These assets are in the `Resources` folder next to the `Finish` folder. As usual, merge these with your project files.

Understanding the camera system

In *Chapter 4, Adjusting Cameras and Lights*, we briefly touched on the concept of a camera in Blender. We learned that we couldn't render a scene without one. Although we took a render in the end by introducing a camera, we never talked about the different settings a camera can have. That was done intentionally because the know-how we would attain in Blender would not directly transfer to Godot. Fortunately, now is the right time to study in detail what a camera can do for enhancing the gameplay experience.

Not only are we going to get to know how to set up a camera that suits our game, but we are also going to discover different types of cameras Godot has in its inventory. As usual, or as it is something you might hear as a joke on internet forums and memes, there must be a node for this type of thing in Godot.

Yes, there is. In fact, there are four camera nodes, as outlined here:

- **Camera**: This is the core class that serves as the base for all the other camera types. Even though you can have multiple **Camera** nodes in your scene, there can only be one active camera. And, similar to Blender, no camera means nothing to see here.

- **InterpolatedCamera**: This is an enhanced version of the **Camera** node. It comes with three extra properties that turn a regular **Camera** node into a mechanism that tracks and follows a target. It's quite handy if you are developing a game with an over-the-shoulder camera look. If the game character is the target, when the target moves, the camera will catch up.

  Unfortunately, this node will be removed in Godot 4. Luckily, it's not difficult to recreate its functionality by attaching a short script to a **Camera** node. In other words, if you remove the fancy bits of an **InterpolatedCamera** node, you get the **Camera** node, hence the decision to drop it in future versions.

- **ClippedCamera**: This is another type of special **Camera** node, and fortunately, it will be kept in Godot 4 since it's an advanced camera system. Our game is currently not using **PhysicsBody** nodes that are responsible for determining which objects can pass through each other or bump and bounce back when the bodies in motion connect with a colliding surface. For that reason, we won't investigate this type of camera, but you might want to check this one out if you don't want your cameras to travel through walls. It will behave like an object respecting physics rules.

- **ARVRCamera**: You might have guessed it: this is used for **virtual reality** (**VR**) projects. It isn't something you'd utilize as a standalone node since it depends on a lot of other nodes that have **augmented reality/virtual reality** (**ARVR**) at the beginning of their names. Building a VR project is an advanced topic that deserves probably a whole book dedicated to the subject. For that reason, we'll skip this node too.

Besides the camera nodes for 3D workflows, there is also the **Camera2D** node that is used in 2D projects. Thus, there are five types of cameras in total.

Out of all these types we presented, the most promising candidate is the **InterpolatedCamera** node. Why? Because an **InterpolatedCamera** node is essentially a **Camera** node with extra pizazz such as target and track functionality. So, in your Godot 3 projects, you can start with **InterpolatedCamera** and treat it like a **Camera** node until you need that extra functionality.

Since we are continuing our work from the previous chapter, it makes sense to tidy up some loose ends. Let's start with that first, then we can move on to introducing camera settings.

Tidying things up for interactivity

The last thing we did in the UI.tscn scene was skinning UI components. During that effort, we had already turned off the visibility of the **Panel** node that was responsible for displaying the note from Clara's uncle. Then, we introduced a series of UI nodes, all grouped under an **HBoxContainer** node. We'll turn that container off too, but let's run the project first by pressing *F5*. You might see something like this:

Figure 12.1 – The first run of our game

The UI decisions we have made are visible in the top-left corner of the game. We don't need those for the moment. So, bring up the UI.tscn if you have it closed, turn off the **HBoxContainer** node, and run the game again. We'll look into some UI concerns in the *Detecting user input* section soon.

Perhaps you've already noticed from the screenshots we have used in previous chapters or simply by looking at the project files that there has already been a Game.tscn scene configured as the main scene for the project. That's why Godot did not ask you to pick the main scene when you pressed *F5* since we had already assigned one to the project for you.

Open Game.tscn, and let's see how this scene is structured. Everything might look self-evident, but there is the root node labeled as **Game**, then two child nodes labeled as **Camera** and **Level-01**. Evidently, the level we created in *Chapter 9*, *Designing the Level*, is a child node in Game.tscn. The other node, **Camera**, will be our main study area in this chapter.

We'll split the rest of our efforts in understanding how cameras work mainly into two distinct areas. The most important topic is the projection type, which fundamentally changes the whole experience. We suggest this be something you decide early on in your own projects since any other tweaking can be done after this choice has been made. So, before we tackle individual camera settings, let's see what kinds of projections there are.

Deciding on a type of projection

If you took an art class on learning how to draw architecture, this might be a topic you are already familiar with. The Godot version we are using comes with three types of projections. Although we will mainly focus on the first two, we'll give a brief definition of all projection types, as follows:

- **Perspective**: This is the default camera projection where the farther objects are from the camera, the smaller they will look. Hence, two objects of the exact same dimensions will look like they are differently sized when one of these objects is placed away from the camera. This is also how human beings perceive the world, so if you don't, get a check-up.

- **Orthogonal**: Also known as **Orthographic**, this type of projection renders objects of the same dimensions without altering their size, regardless of the distance to the camera. This type could give your game the dramatic look it needs. Also, there are some types of games—roleplaying (*Fallout* series) and **Explore, Expand, Exploit, Exterminate** (**4X**) (*Civilization*)—where this kind of projection is preferred.

- **Frustum**: This is a relatively new type of projection that has its uses in some types of games—for example, to get that 2.5D look some old-school games used to have where the visuals looked stretched. If you want to know more, `https://zdoom.org/wiki/Y-shearing` has some information about this topic.

In most cases, the first two projections we listed here will be enough. Maybe it would be better if we investigated their differences by experimenting. Since we've already seen the **Perspective** projection type, it makes sense to try the **Orthogonal** projection type, so follow these next steps:

1. Select the **Camera** node in the Game.tscn scene.

2. Change its **Projection** setting to **Orthogonal** and set its **Size** value to 6.

3. Press *F5* to run the game and notice a different artistic style.

After we make these changes, this is what we have:

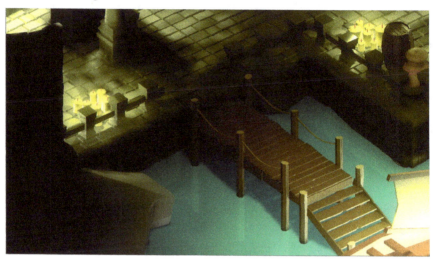

Figure 12.2 – Orthographic camera view from the same location

We picked a **Size** value that would make the render look close enough to the example we had in **Perspective** projection. The **Size** property is an interesting one because it takes into account many factors. For example, if you change the **Keep Aspect** value from **Keep Height** to **Keep Width**, you will have to double the **Size** value to 12. Most PC monitors, however, follow a landscape orientation. That's why **Keep Height** is the default option, but if you are working on a mobile game, you might want to mix and match the correct **Size** value with the **Keep Width** option selected.

Camera-specific environment

While we are looking at different properties of the **Camera** node in the **Inspector** panel, now might be a good time to get a refresher on the **Environment** topic. In the *Creating post-processing effects* section of *Chapter 10*, *Making Things Look Better with Lights and Shadows*, we discovered how to create an environment that changed the look of the level. If you want to override some of the environment settings, you can do so by assigning a separate **Environment** object to the camera. The effects of both the level-wide and camera-specific environments will be combined.

No matter which values you pick for the right platform, one thing is obvious. Even though we didn't move the camera's position and rotation in the world, the effect we get is utterly different. Whereas we used to see the door in the back of the cave in the **Perspective** projection as depicted in *Figure 12.1*, the **Orthographic** view doesn't permit us to see that far, as seen in *Figure 12.2*. When you compare both screenshots, the near elements are pretty much the same, but the **Orthographic** view simulates a more top-down look to the scene than looking far ahead.

Altering stuff in the **Inspector** panel and hitting *F5* to see your changes in effect might get tiring quickly. While the **Camera** node is still selected, if you turn on the **Preview** checkbox, as seen in the following screenshot, you can speed up your workflow when you are editing your camera's attributes:

Figure 12.3 – Previewing what your camera sees is handy, and it's one checkbox away

This will let you preview what your camera is seeing while you are still adjusting its settings. Mind you, during preview, you cannot move around your scene freely. In fact, you can't even select objects. So, remember to turn it off when you want to go back to editing your scene.

In light of what we have presented so far, what kind of projection type should we choose? We're going to go with the **Perspective** mode. So, for now, revert your **Camera** node's **Projection** setting to its default value. Since Godot decorates the **Inspector** panel with the relevant properties, the **Size** property will be replaced with the **Fov** property.

Let's focus on this new property and some of the other changes we want to apply to the **Camera** node in the next section.

Adjusting the camera settings for our game

In this section, we are going to discuss a new term you have just been introduced to, **Fov**, and show which other settings we should apply to the camera. If you have been working on your own level design since the beginning, then the position and rotation of the camera we mention here will be meaningless. That's why we'll give you general directions to convey the spirit of the exercise. Also, hopefully, the screenshots you'll see will help you align our level's conditions to yours better.

First, a quick definition of the new term. **Field of view** (**fov**) is the angle, measured in degrees, through which a device perceives the world. Actually, if you consider your eyes as the device, your eyes also have a fov value. This is a highly technical domain, so we'll offer you a few links in the *Further reading* section to discover it on your own.

For the time being, we're much more interested in the practical applications of this subject since it's pertinent to whether your game is running in portrait or landscape mode, or whether the game is for PC or consoles. The default value, **70**, that Godot uses is a decent average value that will suit most cases. However, this default value also assumes you are going to run your game in landscape mode as it's dictated by the **Keep Aspect** property, which is set to **Keep Height**.

Since players might have different monitor sizes and resolutions, the application has to pick either the height or the width as the **source of truth** (**SOT**) and then apply the other necessary transformations accordingly for the sake of not distorting the visuals. Sometimes, this practice will yield a result such as having a black band above and below the visuals. This method, known as **letterboxing**, is also used in the cinema industry for converting movies shot with a squarer aspect ratio to modern wider (from 4:3 to 16:9 or 16:10 ratio) screens.

If you hover over the **Fov** property in the **Inspector** panel and read the tooltip, you'll see that there are multiple values you can set for this property depending on the aspect ratio your game will use. Thus, we'll let you choose the best value for your condition. Nevertheless, we're providing the following screenshot to demonstrate the permutations of different **Keep Aspect** and **Fov** values:

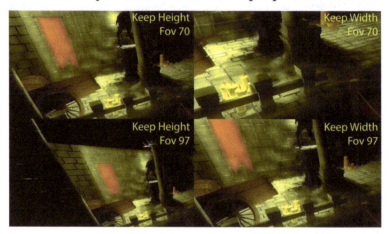

Figure 12.4 – Same camera position with different aspect-ratio constraints and fov values

What a big difference! Without changing a single thing for the camera, different permutations will yield lots of distinct results. Let's wrap up the **Fov** topic by discussing what higher and lower values for **Fov** means so that you can make better decisions in your own projects.

At the end of the day, the **Fov** value you should pick will depend on the player's viewing distance, which isn't something you can really know ahead of time. However, there are conventions you can follow. For example, console games use a lower **Fov** value since it provides a zoomed-in-like view that compensates for the distance between the screen and the player. Most typically, a console game player will be sitting on a couch a few meters away from a screen that is usually large.

On the other hand, a PC player is usually less than one meter away from a monitor, thus it might be better to use higher **Fov** values. This increases immersion since players feel they get to see more of the world by virtue of having this view a bit zoomed out compared to lower **Fov** values. That being said, it's known that really high **Fov** values also create motion sickness. When your brain is forced to process too much of the world, you get that churning stomach feeling, especially in **first-person shooter** (**FPS**) games.

Fov calculator

There is a handy calculator for finding ideal **Fov** values: `https://themetalmuncher.github.io/fov-calc/`. Select the aspect ratio and orientation of your screen, and the calculator will eliminate some of the guesswork. Obviously, if you let your players change their screen resolution in the game's settings, you've got to programmatically update the **Fov** value the game uses.

To finish off this section, we'll stick with the value of `97` for **Fov** and choose **Keep Width** for the aspect ratio since it works out better artistically. Also, since this level is so small, having the camera follow the game character won't be necessary. Still, we could try to pick the best angle and position of the **Camera** node to see most of the scene. As already mentioned, our values won't mean much. However, try to change the **Translation** and **Rotation Degrees** values for the **Camera** node to match what you see here:

Figure 12.5 – The camera's final resting position

What this view will give us are a few things. First, it covers the most crucial angles. Clara can only walk to certain spots on this level. Also, not every walkable location is important. Still, there doesn't seem to be anything significant left out from this perspective.

Second, referring to her uncle's note, there is a backpack behind the broken cart. It's hard to see it from here because the sconce's light in that corner is not enough to make the backpack all that obvious. All of this is intentional because we'll want Clara to hold a torch in her hand, so that extra bit of light will be enough for her or the player to notice an important object.

Eventually, we expect the player to see and interact with the objects in the world, especially the backpack since it holds the key to the upstairs. A common instrument game designers use for player-to-world interaction is mouse events, which is what we'll discover next.

Detecting user input

Mouse events are one of the many types of user input you can detect in a video game. Other most common types are keyboard or game controller events, which won't be covered in this book. Still, the principles in detecting what the mouse is doing are similar to how you can treat other types of events. The reason why we are focusing more on mouse events is that there is an extra layer of complexity you've got to deal with, which is what this section will be about. Let's dive right in.

In a conventional desktop application such as text- or video-editing software, the interface is usually populated with a lot of buttons, menus, and likewise. The natural behavior you'd expect from the users of such software is to click these designated spots, which is something the creators of the application anticipate and prepare for you. How would you go about this in a 3D game, though?

See, when you click anywhere on your screen, you are essentially clicking on a 2D surface. Thus, it originally makes sense to define the click's coordinates based on the x and y axes. Let's make the case even simpler. We are not clicking anything fancy but just the middle of the screen. By knowing the monitor's resolution, we can do the calculation and come up with coordinates that are half the resolution in both axes.

Let's imagine, in this special case where we keep clicking right in the middle of the screen, we have the game world we see in *Figure 12.5*. Where does that click correspond in our level? Even more interestingly, if you implemented a camera that moved elsewhere, perhaps even rotated due to gameplay reasons, how do you map the same x and y coordinates to a different position in the 3D space?

This is a challenging topic that is not always straightforward to resolve, but let's see which techniques we can use to discern mouse events.

Knowing where the player interacts

There is a common technique in the industry for detecting where the player is pointing in a 3D world. It's called **raycasting**, and YouTube is awash with tutorials dedicated to this particular topic, not just for Godot Engine but for other game engines as well. It assumes that you are casting a ray from where you clicked on your screen to a position in the 3D world. Since the game engine is already capable of rendering the game by considering the game objects' positions in relation to the camera, which happens to be your screen, then the calculations are already done for you, to a certain extent.

Although this technique puts you in the right direction, you still have no idea which object in the path of that ray is the one you want to select. Perhaps an unfortunate analogy for a ray might be a strong enough bullet that's traversing through all objects it connects with. So, if raycasting brings up many results, you've got to eliminate the ones you don't want. Fortunately, there is a more direct way.

It would be convenient to only assign detection logic to the objects we want. For example, we can introduce a new model to our scene—a parchment, to be specific—right on the wooden slats of the pier. Once the player clicks this object, we'll trigger the note currently hidden in the UI.tscn scene. Via this effort, you will also practice some of the methods you used in earlier chapters too. Here are the steps to take:

1. Make a new scene out of `Parchment.glb` and save it as `Parchment.tscn` in the same folder.

2. Since there is a default environment in effect, the scene will be dark, and it will be hard to follow the succeeding steps. To disable it, open **Project Settings** and clear the **Default Environment** field in the **Environment** section under the **Rendering** header. Close **Project Settings** to go back to `Parchment.tscn`.

3. Add a **StaticBody** node under the root node.

4. Add a **CollisionShape** node under this last node you introduced and assign a **New BoxShape** to its **Shape** field in the **Inspector** panel.

5. Expand this new shape by clicking it. Type `0.15`, `0.14`, and `0.06` in the **Extents** section's **X**, **Y**, and **Z** fields respectively. This shape should encapsulate the model.

6. Still for the **CollisionShape** node, expand its **Transform** header, then type `0.05` in the **Z** field under its **Translation** section.

We are not done yet with the parchment scene, but let's take a break and explain what's happened.

We have added our first **PhysicsBody** type of node to our workflow with a **StaticBody** node. There are other types too, such as **KinematicBody**, **RigidBody**, and likewise, if you would like to offer physics-based gameplay. Since the parchment object we will place in the world won't go anywhere, we chose **StaticBody**.

Then, we assigned a collision shape to the **StaticBody** node. Adding collision to game objects is necessary if you want the engine to detect when your objects collide with each other. By doing so, the game engine can determine these objects' future trajectory and speed.

One type of collision the game engine can detect is when players interact with objects using input devices. For instance, the player might move the mouse over an object, click this object, or even want to drag and move it somewhere else. Out of all these possibilities, we are only interested in detecting when the player clicks the parchment model. We'll learn how to distinguish the exact event we want in the next section.

Distinguishing useful mouse events

We've constructed all the necessary mechanisms to start detecting collisions. The basic shape we wrapped the parchment model in will act like a sensor to know if collisions are occurring. Out of so many different types of collisions, we are mainly interested in listening to mouse events, and—more specifically—detecting mouse clicks.

We'll treat this click on the parchment as a precursor to bringing up the currently hidden **Panel** node inside the UI.tscn scene. Ultimately, we will build a communication line between the parchment and the UI.tscn scene. First, let's see how we capture a collision and filter out the right type so that we can later trigger the chain of events we want. Here's what to do:

1. Attach a script to the root node in `Parchment.tscn` and save it as `Parchment.gd`.

2. Select the **StaticBody** node and turn on the **Node** panel.

3. Double-click the **input_event** entry under the **CollisionObject** header.

4. Press the **Connect** button in the pop-up menu. This will add a few lines of temporary code, so change the `Parchment.gd` script to what you see here:

```
extends Spatial


signal show_note


func _on_StaticBody_input_event(camera, event, position,
normal, shape_idx):
    if event is InputEventMouseButton and
      event.pressed:
        emit_signal("show_note")
```

We're now, in theory, tracking the input event on the **StaticBody** node. However, in practice, since the collision shape for generating this event is positioned precisely over the parchment, our setup will behave as though you are detecting clicks on the parchment itself. The following screenshot shows our progress in the editor:

Figure 12.6 – We are attaching input events to the parchment object

The input event we are capturing is generic enough, but we are filtering it out so that it will be valid only in mouse-click conditions. Then, we transformed the meaning of this click by emitting a show_note signal, but who is listening to this call? Some construct out there could make sense of this signal—more specifically, the **Panel** node inside the UI.tscn scene. Let's connect them next, as follows:

1. Open UI.tscn and attach a script to the root. Save it as UI.gd and add the following line of code:

    ```
    export (NodePath) onready var note_trigger = get_
    node (note_trigger) as Node
    ```

2. Open Level-01.tscn and create an instance of Parchment.tscn in the **Props** group. Position this new node on the wooden slats of the pier so that it sits relatively close to the boat.

3. Select the **UI** node in the **Scene** panel. There is going to be a **Note Trigger** field for this node in the **Inspector** panel. Press **Assign...** and select **Parchment** among the options that come up in the pop-up menu.

4. Go back to the UI.gd script and add the following lines of code:

    ```
    func _ready():
        note_trigger.connect("show_note", self,
                              "on_show_note")


    func on_show_note():
        $Panel.visible = true
    ```

There is a lot going on here with a few lines of basic code. First, we prepared a field for the **UI** node to accept another object as a trigger so that we could assign the **Parchment** node using the **Inspector** panel. Then, we instructed the **UI** node to listen to a specific event—the show_note signal—so that it could trigger the on_show_note function. When this function runs as a result of the player's click on the parchment, the **Panel** node, which is essentially Bert's note, will become visible.

When you were building the UI in *Chapter 11*, *Creating the User Interface*, if you didn't center the **Panel** perfectly, you can do so now by using the **Layout** button in the header of the 3D viewport. If you prefer, you can position the **Panel** anywhere you want. Ultimately, when you press *F5* and run the game, after you click the parchment on the pier, you will see something like this:

Figure 12.7 – Bert's note to Clara was opened when the player clicked the parchment

Remember that the **Close** button is already wired, so it'll close the note when you press it. If you do so, you can open the note again by clicking the parchment. Who knew that a simple mouse click could mean different things? In one context, it's pressing on a flat surface that translates to clicking a 3D object, which then triggers other game systems. In another, it's pressing a UI element like a button.

> **Sconces and candles**
>
> If the player is able to click the parchment, can't they click the sconces and candles around the level? They can, but they won't get a reaction out of it right now since you have to construct a collision structure, just as we did for the parchment. This is something you can work on as an exercise.

We're not planning to have an inventory system in our game. However, in games that employ that kind of functionality, it's common to see that parchment disappear from the world and find a place for itself in the player's inventory. Then, the player can later click an icon that represents the note in their inventory to bring up the note UI again. In this extra case, your UI structure would also have to listen to a show_note signal emitted from a different structure, but it's a similar principle.

Not having an inventory system is not a real detriment to our workflow at this point since we have more pressing issues such as helping the player move around. Although we have a level where there is a solid floor, we have no game character that can stand on it. We'll look at how to introduce one and move it in the upcoming section.

Moving the player around

You might have heard that context is important in real life because context can make an ordinary word or statement look especially bad or fun. This is consistently true in most technical areas—more specifically when we try to describe visual or artistic aspects. Sometimes, it's alright to use words interchangeably, but making a distinction might be crucial—even necessary every now and then. For example, at the end of the last section, we claimed that we'd move a character. It might be an absurd attempt to do mind-reading via the pages of a book, but would we be wrong if you imagined a biped creature such as Clara walking around using her legs and swinging her arms?

Chances are you did think about it that way, but you'll have to wait for that at this moment since we haven't even moved an object between two spots on the level. Referring to the analogy of context, not every move has to involve a fully-fledged animation. Clara's model, or an ordinary cube for that matter, could also move by following a path. Therefore, it might be more appropriate to think of movement and animation as two distinct topics. That's why we will introduce animation into moving objects later in the *Triggering animations* section after we first tackle movement in this section.

Now that you know there is a difference between an object traversing a scene and doing so with an animation, the big question is: *How to detect where to move an object?* Let's be more specific in terms of our level design. We have a pier where we have just recently placed a parchment. The basic expectation is that our player character will be standing right by this parchment. Once the player is done reading the note, we expect them to reach the backpack to acquire a key to unlock the door that leads upstairs. Therefore, we need a mechanism to do the following:

- Detect clicks
- Find a possible path
- Move the player to their desired spot

Before we can start working on these items, we first need two vital ingredients: **Navigation** and **NavigationMeshInstance**. These two nodes work hand in hand to designate some areas in the level to be walkable. After all, we wouldn't want the player to walk everywhere or through objects, hence the importance of some of the props we placed around the level.

Interchangeability for the sake of brevity

Although we've pointed out a major difference between movement and animation and claimed that we can't use these two concepts interchangeably, we are in luck when it comes to the two nodes we are going to peruse in this section. You'll soon see that a **Navigation** node is practically incapable of doing its work without depending on **NavigationMeshInstance**. We'll use **Navigation** as a general concept (unless otherwise specified) to talk about navigation, while technically, we might be describing the attributes of the **NavigationMeshInstance** node.

With that said, let's create areas that are traversable by the player.

Creating walkable areas with a Navigation node

The level we started to design in *Chapter 9, Designing the Level* has some nice, but also troubling features. From a visual perspective, the props and their placement in the world look organic. Even bulkier objects such as the broken cart and the stag statue are out of the way but still in the line of sight when a person walks between the pier and the door. There is an element of usefulness mixed with clutter.

Speaking of clutter, when we introduce a **Navigation** node and ask Godot to calculate traversable areas, the location of the objects in your level may gain bigger importance. You may get a hint as to why this is after we make changes to the level, so open Level-01.tscn and follow along with these steps:

1. Add a **Navigation** node in the root node. Then, add a **NavigationMeshInstance** node right under this last node you've introduced.

2. Drag and drop the **Floor**, **Columns**, **Rails**, **Props**, and **Dock** groups under the **NavigationMeshInstance** node.

3. Select the **NavigationMeshInstance** node and assign a **New NavigationMesh** to its **Navmesh** field in the **Inspector** panel.

4. Click and expand this new property so that you can do the following:

    I.   Type 0.18 in the **Size** field and 0.1 in the **Height** field under its **Cell** section.

    II. Type 0.4 in the **Radius** field and 0.2 in the **Max Climb** field under its **Agent** section.

    III. Turn on the **Ledge Spans** option under its **Filter** section.

5. Press *7* on your numeric keypad to switch to the **Top Orthogonal** view.

6. Press the **Bake NavMesh** button at the top part of the 3D viewport.

If your level design is different than ours, please try to follow the steps we have presented in the spirit they are given. This is especially important if you directly transfer our values to your system, which might not fit. In the end, you'll see something similar to this:

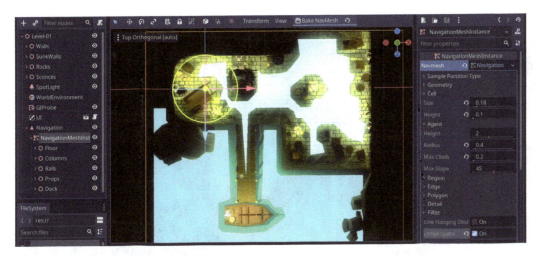

Figure 12.8 – We have introduced a NavigationMeshInstance node and configured it

Notice the light-blue overlay introduced by the **Navigation** node. That is all walkable as far as the engine is concerned. There is something awkward going on, though. When you dragged the **Dock** group into the **Navigation** node, the **Water** node came with it. So, it was also considered a candidate.

If this were a *Dungeons & Dragons* game, your player might know the *Water Walk* spell and be able to walk on the water mesh. There is no such spell in Clara's world, but it's something you might want to consider if your game allows for such a mechanism and flavor. Therefore, instead of removing the water altogether, it's best if we changed its place in the hierarchy by doing the following:

1. Move the **Water** node somewhere other than the **NavigationMeshInstance** node—for example, above the **SpotLight** node.

2. Similarly, drag and drop **Parchment** out of the **Props** group.

3. Select the **NavigationMeshInstance** node and press the **Bake NavMesh** button again.

With a different hierarchy, the newly baked traversable area should look like this:

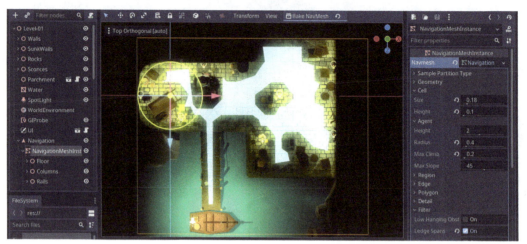

Figure 12.9 – The water is no longer walkable thanks to being in a different hierarchy

By determining which areas should be included in the **NavigationMeshInstance** node and adjusting values in the **Inspector** panel, you can come up with a more precise layout. Ultimately, if you can throw a few obstacles in the player's way before they reach important places instead of following a perfectly straight line, you will create more engaging gameplay.

If the layout in your level doesn't look traversable in some key areas, such as the backpack near the cart, then move some of those props around and bake a new map. This is going to be important when we introduce movement logic.

You might want to rotate the view to **Perspective** if you want to get a better feeling of which areas are reachable. Speaking of which, who is going to walk these areas? Next, we should introduce the most basic player character before we get into more advanced character models such as Clara.

Introducing a basic player character

Earlier in this chapter, in the *Knowing where the player interacts* section when we were inquiring about how the player could interact with the parchment, we introduced a **StaticBody** node because the object wasn't going anywhere. We also mentioned that **StaticBody** was one of many **PhysicsBody** options available to you besides two other commonly used nodes, as described here:

- **RigidBody**: Bodies that don't have control over themselves fall under this category. The word *rigid* might be confusing at first since it conveys a feeling of how strong or flexible an object is. On the contrary, you can use a **RigidBody** node for simulating the motion of a soccer ball or a cannonball. You usually apply forces to objects that have this node, which will instruct how the physics engine will calculate their trajectory, collisions, and likewise.

- **KinematicBody**: Bodies that actually have control over how they will behave in the world fall into this category. Most typically, player characters use this node, but any system that creates its own motion—such as an actual engine or rocket—needs to use this.

Consequently, the best option we have is to use a **KinematicBody** node to simulate a player character. We'll now follow the next steps to create a very simple one:

1. Create a new scene and save it as `Player.tscn` under the `Scenes` folder.

2. Start with a **KinematicBody** node as its root. Then, add a **CollisionShape** node and a **MeshInstance** node under the root.

3. Select the **MeshInstance** node and do the following:

    I.   Assign a **New CapsuleMesh** to its **Mesh** field. Expand this new field and type `0.4` for its **Radius** property.

    II.   Type `90` in the **X** field in **Rotation Degrees** under the **Transform** section.

4. Select the **CollisionShape** node and do the following:

    I.   Assign a **New CapsuleShape** to its **Shape** field. Expand this new field and type `0.4` for its **Radius** property.

    II.   Type `90` in the **X** field in **Rotation Degrees** under the **Transform** section.

5. Select the **KinematicBody** node and type `0.9` in the **Y** field in **Translation** under the **Transform** section. Rename this **KinematicBody** node `Player`.

This will create a capsule shape, which is a quick way to simulate player characters. Also, we picked a collision shape that would go well with the mesh we created. Since there isn't much to look at in the `Player.tscn` scene, it may be best if we show you where to place it in the world. Create an instance of it in `Level-01.tscn`, and position it as shown in the following screenshot:

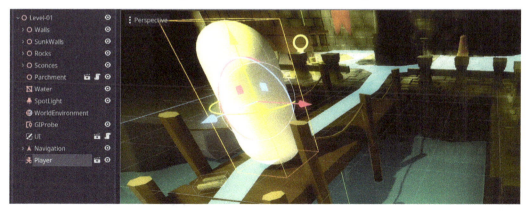

Figure 12.10 – An upright pill-shaped player character

The player character, although it looks like a pill standing up right now, is now part of the world and ready to move around. It just needs to be told where to go. How can we give it instructions before even we know where it's supposed to go? To solve this mystery, we will have to prepare a structure to catch clicks. All this will eventually lead us to revisit a topic we dismissed earlier in the *Detecting user input* section: raycasting. After all, it will help us know where the player clicked in the world.

Preparing a clickable area for raycasting

When you know exactly which objects should be interactive and receive mouse events, the method we applied in the *Distinguishing useful mouse events* section is still valid. It entails anticipation on the game designer's end, so the essential bindings could be done early on, as we saw. However, what if it wasn't always possible to foresee this, or how viable would that method be on a larger scale?

For example, if we were to add a **StaticBody** node to each floor model we have used so far, we could certainly detect mouse clicks. That being said, sometimes, it's a bit too late for that. Right now, our level has all the floor pieces as model instances instead of scene instances because, back then, it was convenient to drop the models and be done with the level design. We could still try to create a scene out of a floor model, but you'd still have to swap all the floor assets in the level. It's a lot of work.

Since we already know that a **StaticBody** node is necessary to initiate an input response, we may yet use it to our advantage. Instead of attaching it to every single floor piece, we could designate an area as large as what all the floor pieces occupy, and detect the clicks on this large piece. Here's how to do this:

1. Add a **StaticBody** node to the level and place a **CollisionShape** node inside this **StaticBody** node.

2. Assign a **New BoxShape** to the **Shape** field in the **Inspector** panel.

3. Expand this new property and adjust its **Extents** setting. We used values such as 9, 1, and 8 but you might want to adjust these values after you finish the next step.

4. Position the **StaticBody** node in the level so that the following applies:

    I.   Its **Y** coordinate is roughly -1.05. Adjust it to a value so that its top almost aligns with the floor but just below the parchment. We'll discuss this after we finish moving the player.

    II. Its **X** and **Z** values are at a point where its child, **CollisionShape**, encompasses the floor pieces and the walkable areas on the pier.

It might be easier to decide on the measurements if you switch to the **Top Orthographic** view. The blue square in the following screenshot represents the area we want to use as a click detector:

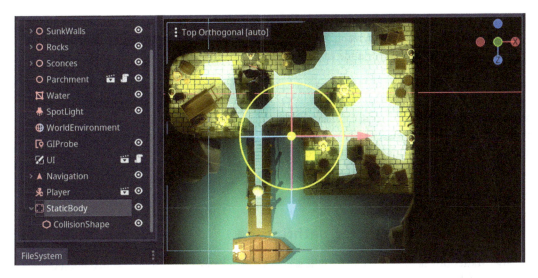

Figure 12.11 – The StaticBody node covers all walkable areas

You might be wondering if we overdid it with the detection area since *Figure 12.11* clearly shows it is way larger than the walkable areas. A short explanation is that when you click on areas outside the traversable field, the pathfinding algorithm will take the player to a nearby spot but never to the exact position the player clicked. For example, if you click in the water, then the player character will move to the clicked spot as close as possible but still stay within the limits.

When you get to see the code, things might make more sense from a technical point of view. With that said, let's attach some code to the player character so that it can move around, as follows:

1. Open `Player.tscn` and select the root node.
2. Attach `Player.gd` from the `Scripts` folder to the **Script** field in the **Inspector** panel.

Let's explain the most important parts of the code we have just applied. You can refer to this code block at `https://github.com/PacktPublishing/Game-Development-with-Blender-and-Godot/blob/main/Chapter%2012/Resources/Scripts/Player.gd`. The first 10 lines are for storing some of the startup values and structures we are going to use. Three of those variables are worth a thorough explanation since the rest is self-explanatory. Let's look at them in more detail here:

- `camera`: The player scene has no **Camera** node, but it needs to access a camera to do the raycasting. So, we appropriate the currently used camera as a workaround.

- `space_state`: This is our entryway to Godot's **PhysicsServer** node that monitors which objects collide or intersect with each other. We're going to use this variable to know if a click connects with the floor.

- nav: Since the **Player** node will be part of the Level-01.tscn scene that also holds the **Navigation** node, we use a mechanism like this to inject the **Navigation** node into the **Player** node. This way, the **Player** node can query the **Navigation** node to find a possible path.

The rest of the script consists of four functions. Despite that, two of those functions are doing the heavy lifting because the _input and _physics_process methods are essentially offloading their tasks to two other functions: find_path and move_along respectively. We could have ignored these latter functions, but when you are able to separate distinct functionality into their own functions, you should do this to keep your code clean.

All of this was done so that we could do a raycasting that is implemented in the find_path function, which is what we are going to study next.

Using Navigation node for pathfinding

The large **StaticBody** node we've added to the scene is still not enough to know at which point on the floor the click happened. Having just that will only let us know that the player clicked somewhere in that area. So, in the end, we are still going to use raycasting for finding the precise location so that we can begin constructing a path toward this position.

To that end, the find_path function in the Player.gd script is going to use the following two techniques:

- First is raycasting, to know exactly where the player clicked
- Second is whether there is a possible path toward that position

The first three lines of code in the find_path function, as shown here, are what raycasting is about:

```
var from = camera.project_ray_origin(event.position)
var to = from + camera.project_ray_normal(event.position) * 100
var result = space_state.intersect_ray(from, to)
```

Firstly, we ask the camera system to tell us from where the ray is going to originate. Hence, we store it in the from variable. This happens to be where the mouse event happened. Keep in mind, though that this event is still on our monitor's 2D surface. There is still no notion of where we are clicking in the 3D world.

Secondly, we ask the camera system to let us know where a ray would go if we projected it 100 units from into the world. Now, we know where to stretch the ray. Still, there is no guarantee that this ray will hit anything. Thus, we check if anything is intersecting the ray, and store it in the result variable.

So, in just three lines of code, we determined a line between where we clicked on our screen and a position in the world. The result of this raycasting might be empty, so it would be prudent to check if there is an object colliding with our ray. Only then can we proceed with finding a path.

This is where the `nav` variable comes into play. Since it's a reference to the **Navigation** node in the level that knows the player's position and where the player wants to go next, it calculates a simple path between these two spots. Ultimately, a series of 3D coordinates are stored in the `path` array.

Separation issues

In a situation such as the pathfinding operation requiring a raycasting done in the `find_path` function—in other words, when two systems are closely related to each other—it might be okay not to separate the raycasting logic into its own function. We'll revisit this concept later when we work on a more advanced game character in the *Triggering animations* section.

Sooner or later, you'll have a walkable path, although this doesn't automatically make the player character follow a path. We'll need several more lines of code to do that.

Moving the player to their desired spot

We have used raycasting to detect a spot where the player wants to go and queried the **Navigation** node to find the closest path to this desired spot. We are now ready to instruct the **Player** node to move between different points along the path.

The `move_along` function in the `Player.gd` script receives a path and processes it one step at a time. Since it's unlikely to have a straightforward path between the start and end points, the path will be composed of a series of midpoints before the player reaches their last stop. It's like walking in real life where you make course corrections before you arrive at your destination. Naturally, if the path is empty or all of its steps have been processed, we terminate the function early.

Otherwise, we move the player between two stops by checking if the distance to the next step is within a certain threshold. Speaking of this threshold, this might be a good moment to talk about a caveat. During the writing and testing of this code, we had moments where the threshold value should have been 3, or sometimes, 1. You might want to experiment with a different value if you notice the player character is behaving awkwardly. This is something that will be remedied in later versions of Godot, as is noted in the official documentation:

> *The current navigation system has many known issues and will not always return optimal paths as expected. These issues will be fixed in Godot 4.0.*

After all this hard work, we are now one step away from having the player character move around, so let's carry on with this, as follows:

1. Switch to `Level-01.tscn` and select the **Player** node.

2. Using the **Inspector** panel, click the **Assign...** button in its **Nav** field to select the **Navigation** node in the upcoming pop-up screen.

3. Press *F5* and click on different spots in the level.

When we test the scene and move the character away from the pier, this is what it looks like:

Figure 12.12 – The player character can now move in the world

You now must be able to move the player character around by pressing on the floor or even in the water. The nearest spot will be picked as a destination. Also, while you are moving around, try to click the parchment on the pier. If it is placed just so it's below the catch-all **StaticBody** node, then you won't be able to trigger the note. If that's the case, either adjust the **Y** position you set in the *Preparing a clickable area for raycasting* section for the **StaticBody** node or move the **Parchment** node up in the **Y** direction.

As long as the clicks are not competing, the parchment will trigger the note. If the player character is away, it will then move near the parchment as soon as the note is open. You might notice odd behavior at this point if you click the **Close** button. The note will close as expected, but the player character will suddenly move just under where the **Close** button was. It's as if the note UI is letting some of our clicks through and the pathfinding logic picks up that call.

Fortunately, there is a quick fix for this kind of behavior. If you replace the `_input` function with `_unhandled_input`, then all will be well. If these two look alike and unclear, you might want to find their nuances in the manual: `https://docs.godotengine.org/en/3.4/classes/class_node.html`. It might be worth remembering its use for quickly fixing a lot of UI headaches.

Wrapping up

If you have been developing video games for a while, you might already be familiar with the notion of iterative and incremental workflow. For example, it's been okay to have indestructible crates so far. Let's examine a scenario where you now want these crates to be destructible.

Not only do you have to account for certain conditions to happen, such as if the player has the right item to break the said crates, but you will also have to prepare animations to be triggered at the moment of destruction. These are both programmatical and artistic changes, and they can definitely be done with ease to a certain extent. When you *baked* the walkable areas, the **Navigation** node believed the crates were solid obstacles. However, in this new dynamic situation, you also have to update the **NavigationMeshInstance** node with the new conditions.

If a crate the player has just destroyed is no longer part of the world, and that particular area should indeed be walkable, you have to update the walkable areas by baking a new map. Fortunately, it's possible to create multiple **NavigationMeshInstance** resources and save them on the disk so that you can swap them to accommodate dynamic cases as needed.

Sometimes, it makes more sense to move ahead with prototypes. For instance, it was good enough to have our player character look like a capsule to test movement logic. It would be nice to have our avatar look more like a person than a white pill. Let's see how we can accomplish that next.

Triggering animations

In *Chapter 5*, *Setting Up Animation and Rigging*, we tackled the creation of animations in Blender. Then, in *Chapter 7*, *Importing Blender Assets into Godot*, we saw how to import a model into Godot Engine and use the **AnimationPlayer** node to test the model's different actions. The steps we'll present in this section should be enough to introduce Clara to the game, but if you need a reminder on how to create and import animations, you might want to seek out those two chapters.

Since we are done with the player's movement, what is missing is to introduce Clara to our workflow and play the proper actions, such as idling while she's standing and walking while she is moving around.

We've already created a basic player character when we constructed `Player.tscn` and attached a script to this scene. It's primitive but the scene structure is a good starting point. Follow these steps:

1. Click `Clara.glb` in **FileSystem**, then bring up the **Import** panel.

2. Select **Files (.anim)** in the **Storage** dropdown under the **Animation** header. Refer to the *Separating actions* section from *Chapter 7*, *Importing Blender Assets into Godot*, to remember the need for this step.

3. Press **Reimport** to set up Clara's dependencies properly. Switch to the **Scene** panel.

4. Open `Player.tscn` and delete the **MeshInstance** node.

5. Drag `Clara.glb` from **FileSystem** onto the **Player** node. Thus, the old **MeshInstance** node will be replaced with a **Clara** node.

6. Click the root node and zero its **Translation** values since the values that applied to the basic capsule-shaped player are no longer valid.

7. Adjust the **CollisionShape** node's **Shape** field in the **Inspector** panel so that it encapsulates Clara. We haven't changed the **Radius** setting but set its **Height** value to `1.2`.

The main goal here is to replace the old **MeshInstance** node with Clara and adjust the **CollisionShape** node so that collision detection is done correctly. The editor should now look like this:

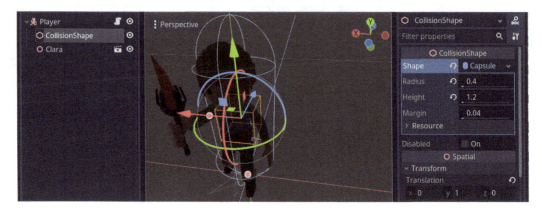

Figure 12.13 – Clara has replaced the boring MeshInstance node

With this method, you can easily test your code for a player character, and then replace the test model with the actual model later. This might be useful if you are the main developer and you are still waiting for the artwork from your colleagues.

Improving the looks of the avatar was a good step forward. It's looking much more appealing. We will do the same for its movement because you might notice odd behavior if you run the game now. Clara will be moving around like the old capsule mesh because it's missing two major qualities, as follows:

- Looking in the direction it's currently moving
- Showing signs of walking instead of looking like a stick sliding on surfaces

There is also another problem but it's so minor you can fix it without needing much discussion and explanation. The **Player** node, which used to hold the simple **MeshInstance** node, had to be moved slightly higher in the world. You can lower this new **Player** node to the level of the pier so that Clara's feet are connecting with it. If you don't make any changes, Clara will look like she's hovering and then moving diagonally as soon as her movement logic kicks in.

For the other two major concerns, we'll have to dig deeper than just changing an object's position. We've got to first update the script we are using for the **Player** node, though, so here's what we need to do:

1. Select the root node in `Player.tscn`.
2. Swap its script with `Clara.gd` from the `Scripts` folder.
3. Press *F5* and enjoy seeing Clara walking around as a normal person should.

Rejoice—she's walking!

How did it happen so quickly? We will devote the rest of this section to discovering which changes the `Player.gd` script has received to accommodate the new behavior we are experiencing and—undoubtedly—enjoying.

Understanding how Clara looks around

An incremental and iterative workflow is the short and non-technical answer to understanding how Clara looks around, and it's something we advise you to keep in mind when tasks seem monumentally big at first. For example, we were initially concerned with basic movement, which was achieved within the `Player.gd` script. At some point, when you know basic test systems are working, it's time to take things to the next level. That's what happened with the `Clara.gd` script.

We'll now explain the steps we have taken to turn the basic sliding movement into a more elaborate walking animation. As far as having new variables is concerned, we are using a simple flag: `is_moving`. We keep track of this flag in order to understand whether Clara is moving or not. The use of this new variable will soon be discussed in the context of some other changes we have made.

> **New term – flag**
>
> In the programming world, a flag is a variable that means a certain condition has been satisfied. It's often used to determine a system's behavior, like an electric switch with a false/true or off/ on states, hence they are often called **Boolean** flags. However, it is possible for a flag to have different kinds of predetermined values.

A natural behavior for Clara would be to look in the direction the mouse cursor is. Let us remind you, once more, that although the cursor is moving over our monitor's 2D surface, we need to do essential projections into the 3D space to find the proper direction. We were already doing that in the `find_path` function inside the `Player.gd` script. Since we now want a similar raycasting done for determining where Clara is supposed to look, we extracted those common lines from `find_path` to its own function, `get_destination`.

Hence, the more common uses and repetitions you can find in your code, the better it is to separate them into their own functions. This was something we intentionally ignored in the `Player.gd` case for simplicity's sake. However, we now have both the `find_path` and `turn_to` functions depending on `get_destination`.

Just as `find_path` is piggybacking on the `_unhandled_input` function, the `turn_to` function is also using the same mouse `event`. Speaking of the `turn_to` function, let's take a closer look at it here:

```
func turn_to(event):
    if is_moving:
```

```
        return

    var direction:Vector3 = get_destination(event) *
    Vector3(1,0,1) + Vector3(0, global_transform.origin.y,
    0)


    look_at(direction, Vector3.UP)
```

First of all, although we haven't yet seen where the `moving` flag is set, if Clara is moving, we wouldn't want her to keep looking around. So, we have an early `return` statement to terminate the turning behavior. Then, once we determine a suitable direction via the `get_destination` function, we trigger Godot's built-in `look_at` method.

The logic is simple, but the math to determine the `direction` vector in `turn_to` might need a bit more explanation. Normally, the value from `get_destination` would have been enough, but we seem to be multiplying the return value with another vector and then adding it to another vector. This is because the destination given by `get_destination` also includes the *y* axis in the 3D space. We want Clara to keep her posture the same; in other words, we don't want her to look up or down. Those two vector operations are required so that she doesn't rotate in an awkward way.

You can see the weird behavior yourself by removing the vector operations and only keeping the `get_destination` function. When you move your mouse cursor near Clara's body, she may suddenly pivot around her feet and sometimes even flip upside down or sideways. The intricacies due to projections between 2D and 3D are something you'll have to account for in the future, and it's a common occurrence in controlling game characters.

It's nice that Clara is facing where the mouse cursor is. It's also a separate mechanism because she can do so without moving, as you may have already tested with the preceding code block. It would be nice if she kept looking where she was going while walking. This will be done in the enhanced version of the `move_along` function. Let's see how we improved it in this new version.

Adding a looking behavior to moving functionality

It is nice to see Clara looking around while she's standing still, but we will also want her to face the destination she's walking to. For example, if you click near the crates by the wall (more like the right-hand side of the screen), she should walk straight until she clears the pier, then turn and look right, and then keep walking. Similarly, while she's in this new spot, if you click somewhere far away such as near the stag statue or the pier again, she should turn around and walk back in a natural way.

This kind of behavior can easily be added inside the `move_along` function. The way it is, that function already determines how many steps there are left along the path Clara should take. As she's walking toward the point on the path, she may as well look at where she is going. That's why we have a simple `look_at` function call after `move_and_slide` in the `move_along` function.

> **Other useful KinematicBody functions**
>
> We have been using the built-in move_and_slide function of the KinematicBody class. There is a useful function in the same class that might be helpful in levels where the player would like to reach an elevated location by following a slope: move_and_slide_with_snap. Similarly, you might want to check whether the player should perform the next move. If that is the case, the test_move method might be handy.

Also, the fate of is_moving gets decided in the following lines of code:

```
if !path or path_index == path.size():
    is_moving = false
    $Clara/AnimationPlayer.play("Idle")
    return


is_moving = true
```

Notice that, similar to how we do it in Player.gd, the if block checks whether there are steps left along the path. It's exactly at this point we can set the state of the is_moving flag. Consequently, unlike the original version, the new move_along function's if block is making sure the moving logic is turned off when there is no path left for Clara to walk.

If the player clicks a different spot and there is a new path determined, then we turn on the moving flag. As long as there are midpoints for Clara to follow, she'll follow the same steps we've described—face the right direction, walk the necessary distance, face the next direction, walk, rinse, and repeat—until she no longer has any more steps to take.

Besides deciding on the state of the is_moving flag, there is something else going on in that if block in regard to animations. Let's focus on that in the next part.

Playing the right action for Clara

We've already seen how actions are related to animations in the *Separating actions* section of *Chapter 7, Importing Blender Assets into Godot*. They are like what atoms are to molecules. So, when we want to trigger an animation for a model, we actually mean to trigger a particular action. We'll finally utilize this notion and put Clara in action.

We have seen how the improvements we made to the Player.gd script have added extra flavor to Clara's behavior. That being said, she could also benefit from a touch-up in the animation department. That's precisely what's also happening inside the move_along function.

We already know how to determine whether Clara should move or not, and we are keeping track of that with the `is_moving` flag. Subsequently, that's the right moment to trigger the required action for her. Thus, when she's no longer supposed to move, we trigger her **Idle** action. Conversely, the **Walk** action is activated when `is_moving` is set to `true`.

When we made `Clara.glb` part of the `Player.tscn` scene, and it turned into a **Clara** node, an **AnimationPlayer** node already came within it with all of Clara's actions set up. The code we have written so far is aware of exactly where this **AnimationPlayer** node is in the internal structure. Should you import a different model with a different **Scene** tree, then you might have to alter your code to find the right path to the **AnimationPlayer** node.

It's hard to convey an animation via the static pages of a book, but when we move Clara near the column approaching our camera, this is what it looks like:

Figure 12.14 – Clara can now move around the level

Also, notice how the light conditions in the scene are affecting her model when she walks in darker and brighter spots. In the next chapter, we'll turn off some of the light sources. So, as she or the player is walking around, we can use the torch she's holding to illuminate the scene.

We couldn't finish the animation topic without mentioning an advanced subject: blending animations. We will not cover it in great detail, but it is worth mentioning it for more advanced cases of animations you would like to use in your projects.

Blending animations or actions

After a while, the names that are used in most computer-related things may start to seem like they have something to do with each other. We used Blender in the first several chapters to build assets, textures, animations, and likewise. The blending we are now going to talk about has nothing to do with Blender itself.

Our point-and-click adventure game is very simple so far. Clara plays the **Idle** action while she's standing still, and she uses the **Walk** action when she's moving. Although her model contains other actions such as **Death**, **Run**, and likewise, we aren't going to use those. If you want to take this project and move it further, you can accommodate different needs a player may have by improving the `Clara.gd` script and incorporate these other actions.

At some point, when you have a much more complex system where the actions you are triggering come to an end to give room to another action, you may notice that these actions abruptly start and end. Then, imagine how the crossfade functionality in an audio player makes the whole experience more pleasant when a song track changes to another. What if you had a way to transition the end of an action smoothly to the beginning of the next action? You can easily achieve that for animations with the **AnimationTree** node.

Unfortunately, the page count is limited for us to cover such an advanced topic. That being said, the official documentation has a nice and long tutorial page dedicated to this very subject. It also comes with plenty of animated GIFs that you can't possibly experience on the static pages of a book. So, in the end, you might be better off exploring how to blend animations by following the instructions at `https://docs.godotengine.org/en/3.4/tutorials/animation/animation_tree.html`.

Blending animations is heavily used in high-pace action games when transitioning between different attack and run states should look more seamless. In our current situation, we are not missing out much by not having this kind of functionality.

We have made great strides so far in our point-and-click adventure game. Let's wrap up and count our victories.

Summary

This chapter finally covered the long-due camera topic we've been waiting to tackle since *Chapter 4, Adjusting Cameras and Lights*. You now have multiple options to choose from, from a simple **Camera** type to an **InterpolatedCamera** type that follows a target. Should you want to get fancy and dip your toes in VR, you also have **ARVRCamera** at your disposal.

As you now had a proper camera showing you the game world as opposed to seeing things within the editor, it was the right time to investigate how to interact with the world itself. To that end, we presented **raycasting** as a possible solution but quickly dismissed it in favor of using collision detection, which provides more flexibility and precision. We used this technique to detect a click on a specific game object: a parchment. During this effort, you used **signals** as a way of interpreting the player's click as a trigger to turn on the note.

Next, you looked into creating a simple game character and moving it around the level. Sometimes, the game design is missing key elements, and other times, the level might need some more help. Once you, as a solo developer or as a team, are happy with the direction the game is going, then you can up the ante by introducing more complex systems.

That's exactly what happened in between testing `Player.gd` and improving it with the more advanced `Clara.gd` script. In the end, you were able to find a suitable position in the world for Clara to move and do this by using the proper animation cycle. Since you've covered the essentials, it's now up to you to enhance the script if you want to use more actions and special conditions for player interaction with the world.

In the next chapter, we'll introduce a few more tools that will add to the interactivity we have been building, such as playing sounds, conditionally triggering certain events, and switching to another level.

Further reading

Although we taught you how to technically set up a camera, there is a whole other artistic side to picking the best camera settings. You might want to check out online courses and books that cover topics such as composition and storytelling. A few examples are provided here:

- `https://www.udemy.com/course/composition-and-perspective-for-stunning-visual-art/`
- `https://www.learnsquared.com/courses/visual-storytelling`
- `https://www.learnsquared.com/courses/production-concept-art`
- `https://www.cgmasteracademy.com/courses/93-composition-for-concept-art-and-illustration/`

If the code in the `Player.gd` and `Clara.gd` files look very similar, and if it's hard to compare line by line, there is an online tool you can use that can help you see and highlight the differences: `http://www.tareeinternet.com/scripts/comparison-tool/`.

Our game doesn't involve enemy characters that follow our player, but it will follow a similar approach. For example, once the enemy detects the player, it will also have to do pathfinding for finding the player's position and moving toward it. A lot of video-game AI books cover player detection and seeking topics such as the example we gave. So, since most AI topics are generally universal, don't shy away from reading a wide variety of material. You can always apply the insight you gain elsewhere later in your Godot projects.

13

Finishing with Sound and Animation

We're in the home stretch. The effort we started back in *Chapter 9, Designing the Level*, was resumed by making the level look more exciting in *Chapter 10, Making Things Look Better with Lights and Shadows*, which led us to implement a basic user interface in *Chapter 11, Creating the User Interface*. We built new mechanics in *Chapter 12, Interacting with the World through Camera and Character Controllers*, so we could interact with the world we have created. As a result, Clara is now able to press the parchment left by her uncle, and she can also walk around. This is all very nice, and we can take it a step further by refining some rough edges.

It's all quiet in here! As she's walking, we should trigger an audio file that will simulate her footsteps. While we are at it, we will also add background music and effects that will better reflect the qualities of the environment Clara is in.

You must have noticed that, as Clara walks around, sconces and candles around the level illuminate her. Can she do the same with the torch she is holding in her hand? Of course! It might help her see the backpack behind the cart. In fact, she's going to have to use her torch to see better because we'll turn off all of the light sources in this chapter.

We'll discover a new node in Godot to know whether a player character entered an area. Via this method, game designers usually trigger in-game events such as traps, a conversation with a quest giver, and so on. Our event choice will be Clara lighting the sconces and candles as she goes near them.

Eventually, she'll reach the backpack where she'll pick up the key. We are not concerned with an inventory system in this game, yet we will consider this key object as a requirement for opening the door. So, once the condition is satisfied, we need that door to open for us. However, the door did not come into Godot with its animation set up in Blender. This is our chance to see how basic animations can be created inside Godot.

When all of the conditions are in place, including the door opening that simulates a clear path upstairs, we'll swap our current level with another one. That particular moment will signify the conclusion of our little game, but you can take it wherever you want to take it.

This is going to be another chapter with lots of distinct topics used together. Speaking of which, the following are the titles under which you'll find us executing the plan we've presented so far:

- Playing music and sound effects
- Creating reaction spots
- Building simple animations in Godot
- Loading another level

By the end of this chapter, you'll have finished the core mechanics of our point-and-click adventure game. Not only will you construct and work with new systems, but you'll also make these systems conditional on world or character events.

Good luck and enjoy!

Technical requirements

It's perfectly fine if you would like to continue where you left off in the previous chapter. However, there are some extra resources you will need to finish the work in this chapter. You can merge these assets with the rest of your project files. They are in the `Resources` folder next to the `Finish` folder in this book's repository that can be found at `https://github.com/PacktPublishing/Game-Development-with-Blender-and-Godot`.

Playing music and sound effects

Music and sound effects sometimes can make or break the enjoyment people get out of movies, theatre plays, and of course, video games. When done right, they will definitely add to the immersion. In this section, we'll tackle the use of music and sound effects from a technical point of view. In your own free time, we suggest you investigate the artistic aspects of sound design in multimedia for which we'll mention a few resources later on in the *Further reading* section.

In *Chapter 8, Adding Sound Assets*, we discussed different nodes Godot uses to play sound in different dimensions, as follows:

- **AudioStreamPlayer3D** for conveying 3D positional information to the player. It's most commonly used in FPS games where not only front and back directions matter, but an audio stream coming from an elevated place is important as well.

- **AudioStreamPlayer2D** for games in which the direction the sound is coming from doesn't need to have depth information. Most platformer games are a good example of this kind.

- **AudioStreamPlayer** for background music since it may be considered one-dimensional.

Out of these three, two types seem to be the right candidates for our purposes. We want to play background music, so we will use **AudioStreamPlayer**. Then, when Clara is walking around, it makes sense to use **AudioStreamPlayer3D**.

The latter case may not seem obvious, and we can certainly use the regular **AudioStreamPlayer** as well for the footsteps, but we will cross that bridge when we come to it. Our most immediate concern is to set up the ambient music.

Setting background music

In the *Understanding the camera system* section of *Chapter 12, Interacting with the World through Camera and Character Controllers*, we showed the use of an outer scene structure, such as Game.tscn, to hold the level we built in *Chapter 9, Designing the Level*. A wrapper structure such as ours is also a good place to place more global-scale constructs, such as audio streamers. Yet, we would like to discuss an alternative before we move on with our initial plan.

Although a player character is part of the game world, we decided to place it inside the level via a **Player** node. It was convenient to do so because we could easily see where to position **Player** inside the coordinate system of the Level-01.tscn scene. If you place it inside Game.tscn, for the sake of keeping things separate and sanitized, then you will have to figure out a way to connect both the **Player** and the **Level-01** nodes inside Game.tscn. This would not be impossible, but it would make things less convenient.

Similarly, where should you place the node that will play the background music? Although we may want every level to play its own thematic music, and this would guide us in the direction of using an **AudioStreamPlayer** node inside each level, we'll still place it in Game.tscn. When we attack the topic of loading different levels in the *Loading another level* section, hopefully, the scheme we are suggesting will make more sense.

Let's see how we can execute the original plan. Open the Game.tscn scene and perform the following steps:

1. Add an **AudioStreamPlayer** node to the root and rename it as **BackgroundMusic**.

2. Drag Native Dream.mp3 from **FileSystem** to the **Stream** property of this new node.

3. Turn on the **Autoplay** option in the **Inspector** panel.

4. Press *F5* and relax.

The piece of music we are using is about 2 minutes long and it will be automatically looped by Godot. Thus, it won't feel too repetitive while Clara or the player is discovering the level.

Speaking of placing a background music structure at a higher level, there is one more approach you can use: **singletons**, also known as **AutoLoad**. For absolute beginners, these are the ultimate top-level structures you can use in your project. These will always be present when you launch your game and loaded in the order you define them in the **AutoLoad** tab of **Project Settings**. Via this method, you can use a dedicated scene as a single source of music. You can read more about it at `https://docs.godotengine.org/en/3.4/tutorials/scripting/singletons_autoload.html`.

Some players turn off game music for the sake of focusing on sound effects. In the following section, we'll introduce our first sound effect. We expect Clara's walking to trigger a suitable sound effect, namely footsteps.

Conditionally playing a sound

Let's see how we can play a sound file conditionally in this section. There is actually nothing magical nor special in the way of achieving this goal. It's similar to knowing when Clara walks or stands idly. In the *Playing the right action for Clara* section of *Chapter 12, Interacting with the World through Camera and Character Controllers*, we implemented two extra lines of code inside the `move_along` function to trigger the correct actions for Clara to show, animation-wise, the state she is currently in.

We could still take advantage of the same function by enabling the execution of the sound file for her footsteps. That being said, now might be a good moment to discuss some of our practices. It would seem that we are overloading the meaning of the `move_along` function. You might consider our current efforts still a phase of building a prototype similar to, as is often said during a writing exercise, writing a draft, then focusing on edits later.

Sometimes, good architecture might be deduced before you start the bulk of the work, perhaps because you've done something similar before. Often, though, this may not be the case, and your discoveries, thus your decisions into coming up with an efficient architecture, might have to wait for later. As soon as you notice there are common parts you can extract out of the current structures, you should. However, concerning yourself with the fine details of creating the most efficient code structure and information flow might not be the best use of your time while you are still deciding on gameplay.

So, for now, we'll add the footsteps sound as an extra element inside the `move_along` function until we need a much more efficient way, as follows:

1. Open the `Player.tscn` scene and add **AudioStreamPlayer3D** under the root node. Rename it as `FootSteps`.

2. Select `FootSteps.wav` and switch to the **Import** panel. Then do as follows:

    I.   Turn on both the **Loop** and **Normalize** options.

    II.   Press **Reimport**.

3. Drag `Footsteps.wav` from **FileSystem** to the **Stream** field in the **Inspector** panel.

4. Turn on both the **Autoplay** and **Stream Paused** properties.

5. In the `Clara.gd` script, do as follows:

    I.   Type `$FootSteps.stream_paused = false` after you trigger her walk action.

    II.   Type `$FootSteps.stream_paused = true` after you trigger her idle action.

The method we are using here was discussed in the *Playing a sound effect on demand* section of *Chapter 8, Adding Sound Assets*, when repeatedly triggering a sound file in a loop might sound like the sound is jammed.

Additionally, we turned on the loop feature and normalized the volume. The loop is self-explanatory since we will want her footsteps to repeat ceaselessly as long as she's walking. The **Normalize** option deserves a few more words, though. The sound files we are using in this project have been collected from multiple sources. This makes it hard to have all these files have a similar level of volume. Some will be louder, some will be quieter. The feature we turned on adjusts the volume of the sound file, so it would be at a similar level to the other files.

When you run the game now, you'll hear the background music as usual. Then, click around and wait for Clara to walk to the desired spot. Do you hear her footsteps? Most likely just barely. We'll look into adjusting audio volume later in the *Understanding the volume through decibels* section.

For the time being, it might be better if we presented a handy feature in Godot. There might come a time when you would like to apply special effects to some of the sound files you are playing. Godot offers multiple audio channels, also known as an **audio bus**, via which you can decide which files will play on a specific channel so you can apply a particular effect only on select channels.

We'll now pretend that there is a situation like this and play the footsteps sound in its own audio channel. Let's see how it is done as follows:

1. Expand the **Audio** panel at the bottom section of Godot Engine. Click on the **Add Bus** button in the top right corner of the **Audio** panel.

2. Rename this **New Bus** as `SFX`.

3. Select the **FootSteps** node and choose the **SFX** option in the drop-down options for **Bus**.

The footsteps sound will now be played on a different audio channel in Godot. The interface that's reflecting the changes we have made is shown in *Figure 13.1*.

Figure 13.1 – We are playing the sound effect on its own bus

Via this method, a dedicated audio channel will play the sound you want. As you can see at the bottom of the **SFX** bus in the **Audio** panel in *Figure 13.1*, the audio is sent to the **Master** channel. When all the audio sources are merged and processed, it's delivered to **Speakers**. Furthermore, by using the **Add Effect** dropdown for an audio bus, you can apply and stack effects that go through this channel.

Although you hear both pieces of audio, they might be competing volume-wise. In the following section, we'll get a bit technical about how audio volume works.

Understanding the volume through decibels

Every vocation has its trade secrets and unique practices, and this is also true for sound engineers. When they talk about volume as how loud a sound is, they use a unit called **decibel**, marked as **dB**. If you are used to the metric system, this is one-tenth of a bel, similar to a decimeter as one-tenth of a meter. However, what exactly is a bel?

Wikipedia has a page that provides a decent amount of technical information for the decibel. Therefore, we'll provide you with the practical aspects and/or pitfalls of working with decibels in your projects.

Similar to how earthquake magnitude is measured, a decibel is a relative scale where every time you increase the sound level by 6 dB, you double the amplitude of the sound. Consequently, -6 dB means you are halving the amplitude. As far as values go, 0 dB is the maximum amplitude a digital audio system will use. Anything above this value, which means positive values, will be clipped. So, you might still hear something above 0 dB, but it will be distorted the higher you go in decibels. Thus, you'll be using the negative range when it comes to picking values.

Moreover, there are physical limits to human hearing. Sound is no longer audible between -60 dB and -80 dB. So, in the end, you have from -60 dB to 0 dB as a workable range. If all of this is confusing, there is perhaps one important fact you might want to keep in mind about decibels. 0 dB denotes the normal amplitude of the sound when it was exported from an audio application. If the base level at 0 dB is too quiet, you might have to fix it at the source rather than messing with it by choosing a higher dB value in Godot.

That being said, we can decrease the amplitude easily. This is indeed what we are going to do with the background music as follows:

1. Open `Game.tscn` and select the **BackgroundMusic** node.

2. Adjust **Volume Db** in the **Inspector** panel to `-12`, or even `-18`.

Since you are now able to discern the footsteps from the background music better, did you notice how Clara's footsteps get louder as she approaches the camera and quieter as she walks toward the end of the cave? This is thanks to the **AudioStreamPlayer3D** node's behavior of processing audio in 3D. If you want to perceive this effect more clearly, feel free to temporarily turn off the background music and focus on the directionality of Clara's footsteps.

Who is listening?

The **Camera** node has a built-in **Listener** construct that makes it possible for us to identify from which direction the sound is coming. In some cases, we may want the camera to be in one corner of the world and the listener in another corner. Thus, creating a separate **Listener** node is not only possible, but it will also be beneficial when you want to simulate a situation where a microphone is placed away from the camera.

If you would like to practice more on playing sound files, we suggest you add a sound effect to the **Close** button we used in the note user interface. You already know when the button is clicked since we wait for that moment to close the interface. That is the right moment to play a sound effect such as `ButtonPress.wav` in the `Audio` folder.

It seems the world is reacting to our actions by playing animations and sound files, which is nice. In all of these efforts, we've had a direct involvement mainly by a mouse click. In the following section, we'll discover how the world can react to our player character without the player's direct intervention.

Creating reaction spots

When the player clicks on the parchment, the game shows the content written on that parchment via a user interface. When the player clicks on a particular location in the world, Clara walks to that spot by playing a walking animation and playing a footsteps sound. These are all direct interactions at the player's end, which brings us to discuss cases when the game should react to indirect events.

Although not lit, Clara is holding a torch. You already know how to use the **Light** nodes in Godot. So, it's easy to place **OmniLight** near the torch mesh inside the **Clara** node. Our basic expectation is that, when she walks by the candles on the floor and the sconces on the walls, she'll be lighting those up using her torch. Thus, the game needs to know when she's near some objects.

Let's first give Clara a torch she can carry around, then we can proceed to discuss how this torch can affect other objects in the level, as follows:

1. Create a scene out of `Clara.glb` and place an **OmniLight** node under **Torch002**.
2. Position **OmniLight** according to the torch so it's slightly above. `0.75` on the **Y** axis might be enough.
3. Select `d6d58e` for **Color** and turn on **Enable** in the **Shadow** section.

Since **OmniLight** is a child of the torch mesh, whenever the **AnimationPlayer** node controls the torch, the light will follow along. This is also a nice example of taking Blender animations and enhancing them with Godot nodes.

We have a dedicated `Clara.tscn` scene, but the `Player.tscn` scene is still unaware of this new development. It's still using the old model reference. Therefore, you must delete the **Clara** node in `Player.tscn` and instance `Clara.tscn` instead. The **Scene** panel won't look that much different but it's now going to have Clara holding a lit torch. Test your scene and have Clara walk around, especially near the door. The torchlight will synchronize with her walking cycle.

Clara seems to be carrying the right tool in her hand to light those candles and sconces. It's time we added the trigger zones so that the world can react to her presence. That's what's coming up next.

Placing trigger points in the world

We made use of a **StaticBody** node to detect user clicks in the *Preparing a clickable area for raycasting* section of *Chapter 12, Interacting with the World through Camera and Character Controllers*, so we could deduce where to move Clara. This is useful when you know that an agent, most likely the player, will directly trigger a system. There are cases when game objects act freely on their own and they should also initiate a response from systems that are waiting to be triggered. This section will cover this kind of situation.

By now, you may have noticed an odd behavior regarding pathfinding and the player's destination. **StaticBody** that we set up goes as far as where the floor pieces meet the wall pieces. Therefore, it successfully captures the clicks on the floor tiles. However, if you click anywhere far away or along the walls, the pathfinding may give you an unexpected result. If you extend **StaticBody** further out, similar to how it covers the water, it will be alright. You can refer to *Figure 12.11* of *Chapter 12, Interacting with the World through Camera and Character Controllers*, to observe the placement of **StaticBody** and adjust it to account for extra space to catch faraway clicks.

Once the destination is determined, Clara will move toward it by getting closer to the props. Some of these objects are good candidates to trigger certain events. For this, we'll use the **Area** node, which is inheriting from the same internal structure as **StaticBody**. These are similar nodes since they both originate from the same place but provide different results.

Although we could place and position an **Area** node per trigger zone in the level just as we did with many other nodes, keeping in mind that we want to do this for lighting the sconces and candles, it makes more sense to open the dedicated scenes we already have for these. To that end, you will do as follows:

1. Open `Candles_1.tscn` and place an **Area** node under the root.

2. Bring up the **Node** panel and double-click the **body_entered(body: Node)** item.

3. Press the **Connect** button right away, which will automatically add an event handler to the `LightSwitch.gd` script. Change it as follows:

   ```
   func _on_Area_body_entered(body):
       print(body)
   ```

4. Place a **CollisionShape** node under the **Area** node you have just added.

5. Define **New BoxShape** for the **Shape** property in the **Inspector** panel.

The number from the `print` statement might look different in your machine, but you'll see something like **StaticBody:[StaticBody:2025]** in the **Output** panel when you run the game. We've just got a collision result from the **Area** node we've added, but what is it that it hit? It is detecting the catch-all area that covered all of the floor pieces and some portion of the water.

We need to exclude all unwanted candidates so that this trigger zone only responds to our player's activities. There are multiple ways to do this. We'll explain an elaborate version right after we present a very simple method. For now, swap the function you just saw with the following code:

```
func _on_Area_body_entered(body):
    if body.name == "Player":
        print("Hello, Clara!")
```

The changes we are making are in `Candles_1.tscn`, which holds the candle group by the barrel when Clara turns right after she clears the docking area. So, press *F5* to run the game and move her near the candles as described. You'll see the **Output** area display the print message only when she enters the space of those candles. *Figure 13.2* will help you see what's expected.

Figure 13.2 – It's as if the candles sensed Clara coming nearby and welcomed her

With this method, we are only interested in knowing whether name of the body that entered the **Area** node's space is equal to `Player`. If so, we can trigger the next chain of events. However, before we start tackling our initial intentions, the following are a few words about a more advanced detection method we mentioned.

Getting to know a better collision detection method

Godot's **PhysicsServer**, a system that's responsible for undertaking all of the calculations for the objects that should be affected by physical rules (such as gravity, collision, intersection, and so on) uses a layer system to keep track of where objects reside. This is not a visual layer as you might see in a graphics editing application such as Adobe Photoshop. Nevertheless, it's similar because if the objects are on separate layers, then you can define how these layers will interact with each other. Aptly so, the structure that allows this kind of functionality is called **Layer** in Godot.

Moreover, if all objects are always in the same layer, then you would have to resort to solutions such as name checking. It's simple and effective, but it could easily get unwieldy because who would want to pick a unique name for each game object? Unquestionably, that `if` block we wrote earlier would get longer and longer to filter which particular object entered the area. To eliminate such situations, Godot has another construct that is called **Mask**.

Through a clever way of creating multiple **Layer** and **Mask** options, and defining their relationship, you can reduce the load of writing unnecessarily long and inconvenient `if` blocks where you check what's colliding with what. In a way, that sort of check will be done for you in **PhysicsServer**, so you can only account for completely necessary `if` checks for controlling other less trivial cases.

The following figure shows where you can find the **Layer** and **Mask** options for the **Area** node we are currently configuring:

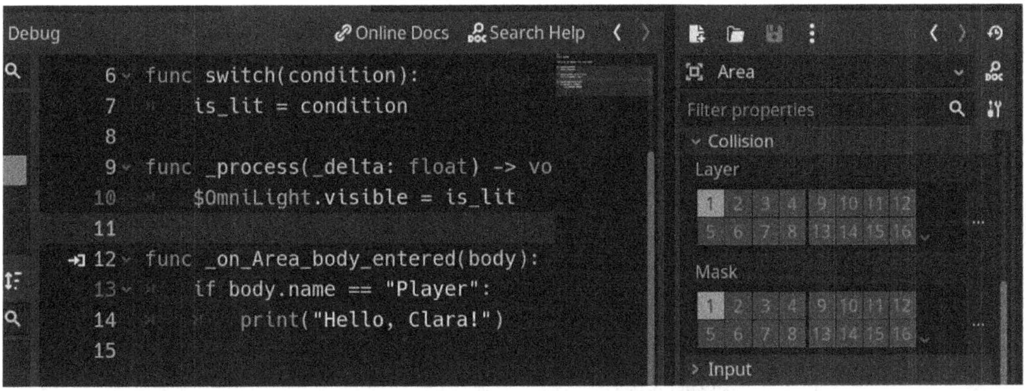

Figure 13.3 – Using collision layers might be another detection method

While this method is effective and valuable, setting it up in our current situation and explaining it via the pages of this book would be inefficient. Instead, we will use the available space to present other practical applications. Still, it is a vital architectural choice you might have to rely on in your future projects. So, we suggest you read about this by visiting the *Collision layers and masks* section at `https://docs.godotengine.org/en/3.4/tutorials/physics/physics_introduction.html`.

Our more immediate concern is what we do when Clara goes near those candles. Let's see her influence on the world.

Lighting the candles and sconces

We've been laying the groundwork for Clara to interact with the world around her. Our latest effort involved proximity detection by `Candles_1.tscn` through the use of an **Area** node. The reaction is not useful at this point since it's just a silly `print` statement, but we are at a good spot to make it more interesting.

To truly appreciate Clara's impact on the world, we should start by turning off some of the lights on the level. Switch to the `Level-01.tscn` scene and perform the following steps:

1. Select all instances of `Candles_1.tscn` and `Candles_2.tscn`.

2. Turn off the **Is Lit** property in the **Inspector** panel.

3. Repeat the first two steps for all sconces in the level.

4. Press *F5* to run the game and move Clara around.

Atmospheric, isn't it? When Clara goes to the same spot that triggered the message in the **Output** panel, the level will look like the following:

Figure 13.4 – Clara is depending on the torch she's holding in her hand

The torch she's holding is enough for her to see where she's going. However, it would be nice to light those candles she's just standing by. We've already done the hard work in `Candles_1.tscn` so all there is left to do is to turn on **OmniLight** internally as follows:

1. Open the `LightSwitch.gd` script.

2. Replace the `print` statement in the `_on_Area_body_entered` function by typing `is_lit = true`. The function will look like the following example after your changes:

```
func _on_Area_body_entered(body):
    if body.name == "Player":
        is_lit = true
```

3. Press *F5* to run the game and move Clara first to the same area, then to a different location.

When Clara goes near the same candles this time, those candles will be lit. It might be a bit difficult to see the effect depending on exactly where she's standing. So, when she walks away from those candles, you'll truly feel her mark on the world, as seen in *Figure 13.5*:

Figure 13.5 – Clara is getting some help from those candles she just lit

This was just one candle game object Clara interacted with. We have another candle scene, `Candles_2.tscn`, and a separate scene for the sconces, `Sconce.tscn`. We could easily replicate what we have done to this point for these other scenes, as follows:

1. Open `Candles_1.tscn` first, then right-click the **Area** node, and select **Copy** in the context menu.

2. Open `Candles_2.tscn` next, then right-click the root node, and select **Paste** in the context menu.

3. Bring up the **Node** panel and then do as follows:

   I. Right-click the **body_entered** item in the list and select the **Disconnect All** option. Press the **OK** button on the upcoming confirmation screen.

   II. Double-click the **body_entered** item in the list. Press the **Connect** button on the upcoming screen.

Normally, we shouldn't have to do the third step. When you copy and paste nodes between scenes, the signals are not transferred. So, we had to manually remove what seemed to be an active signal and rebind it. Luckily, both candle scenes are using the same script and we already have the event handler. That's why we didn't have to write the programming parts. When you transfer nodes between scenes as we did, keep in mind to reconnect the signals. Godot 4 might have a fix for this behavior.

So, run the game and have Clara walk by all of the candles. They will be lit one after another as she gets close, and the following is what you'll experience when she does so:

Figure 13.6 – All of the candles were lit after Clara walked by them

We suggest you apply the same procedure to the `Sconce.tscn` scene. This time around though, alter the **Z** axis of **BoxShape** for the **CollisionShape** node to simulate the extra distance the sconces must have between the walls and Clara. We chose a value of 2, but you might want to adjust it to something that suits your conditions. Alternatively, you could move the whole **Area** node a bit forward to line it up with the two extensions of the sconce that connect to a wall. As long as there is enough area extended out of sconces, Clara will trigger it.

So, where else can you take this idea? A simple case might be to introduce traps or enemies reacting to the player's position. In the case of enemies, they can also take advantage of pathfinding via the same **Navigation** node we placed in the level. Also, it's common, in a case like this, when enemies give up after following the player for a certain period of time. If the distance is not getting any shorter and the player is getting away fast enough, the enemy will usually return to their designated patrol zone instead of trying to catch up with the player.

We aren't going to introduce such mechanics in this game. However, it might be something you can pursue as a more advanced game feature. If you are really interested in enemy versus player behavior, then we suggest you read a few **artificial intelligence** books on game development. There are a plethora of options out there and we'll give you a brief list in the *Further reading* section.

There are two more trigger zones we should create. One is for the backpack behind the cart when Clara goes near that area. The other one is when she approaches the door that leads upstairs. Let's start with the backpack.

Adding the trigger for the backpack

This effort will be similar to the way we did it for the candles and sconces. Since you already know that by using an **Area** node you can introduce interactivity, we'll present something slightly new.

When players interact with the world, more specifically with the game objects, they feel that they have agency over these items. For example, players have just discovered that walking near candles will light them. This is part of the fun besides the narrative and story elements a game can have. At this point, it's up to the game designer to interweave another layer of complexity. Perhaps, being close to the candles is only a precondition and the player is also expected to click on the candles.

Regardless of the conditions a game designer will expect the player to satisfy, giving feedback to the player is quintessential. When players try things on their own, they will get negative or positive feedback. This kind of harmless trial and error could easily be used in lieu of a tutorial. An easy and reliable way to provide feedback is something we've already looked at. It is playing sound.

For the backpack exercise, we'll combine both playing an audio file and reacting to an area effect. Once Clara approaches the backpack as she did with the candles, the backpack will play a sound file that will inform the player that she picked up the key. The following steps show you how you do it:

1. Create a scene out of `Backpack.glb` and save it as `Backpack.tscn` in its original folder.

2. Place an **AudioStreamPlayer** node under the root. Assign `CollectItem.wav` to its **Stream** field.

3. Add an **Area** node with **CollisionShape** under the root, similar to how you did it for the candles. Position it at `-2` on both the **X** and **Z** axes. You may want to pick values that make sense in your scene. As long as there is ample room for Clara to reach this zone, things should be fine. Use *Figure 13.7* as a reference.

4. Create a `Backpack.gd` script for the root node and save it in the same folder. Activate the **body_entered** signal for the **Area** node, which will add a boilerplate function to the script. Then, change the script as follows:

```
extends Spatial

signal key_collected

func _on_Area_body_entered(body):
    if body.name == "Player":
        $AudioStreamPlayer.play()
        emit_signal("key_collected")
```

5. Swap the **Backpack** node in `Level-01.tscn` with an instance of `Backpack.tscn`.

We are following the same principles we used in player detection for the candles. This time, instead of enabling lights, we are playing a short sound effect. We chose the **AudioStreamPlayer** node instead of its 3D version because we don't want this sound effect to be affected by its distance to the camera. However, this is a perfect situation for you to swap and try both to see the difference.

The sound effect command is followed by the emission of a custom signal. In simple terms, we have converted the **body_entered** signal into a **key_collected** signal, which will be used in a more advanced scenario in the *Playing the door animation on a condition* section.

As mentioned in the third step, *Figure 13.7* shows the relative position of the **Area** node.

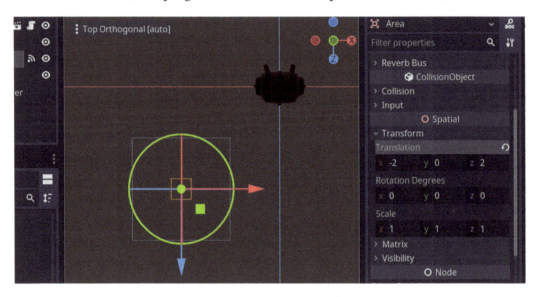

Figure 13.7 – The trigger area for the backpack is offset so Clara can reach it

As they are now, the sconces and candles don't play a sound effect when they are lit. This might be a short and nice exercise for which you can use the `TorchWhoosh.ogg` file. By default, the file's **Loop** feature will be on. So, remember to press the **Reimport** button after you turn the loop off in the **Import** panel.

Last on the list of making some of the game objects interactive is the arched door. Our workflow will be similar but additionally accounts for that `key_collected` signal we defined in this section.

Interacting with the door

You've been using the **Area** node quite liberally for a while. So, you must be used to it by now. In this section, you will use it one last time to complete the topic of interactivity. It will be for the door where you'll also make use of that custom signal we have recently created.

Since some of the steps will be so similar, we will give shorter instructions for the sake of focusing on the unique parts, as follows:

1. Create a scene out of `Doors_RoundArch.glb` and save it in its original folder.

2. Attach the `Doors_RoundArch.gd` script from the `Scripts` folder to the root node.

3. Add two **AudioStreamPlayer** nodes under the root. Rename them as `LockFiddling` and `OpenDoor`. For these two nodes, use `LockFiddling.wav` and `OpenDoor.wav`, respectively, for their **Stream** property.

4. Add an **Area** node to the root with its dependencies and requirements, such as its collision, signal, and position. *Figure 13.8* should be helpful to show where we are placing **Area**.

5. Swap the existing door asset in the `Level-01.tscn` scene with this new scene. Also, assign the backpack asset to the **Backpack** property in **Inspector**.

6. Press *F5* and have Clara walk directly to the door.

We'll pay closer attention to the script this new scene is using after you see how things look in the editor with our most recent changes.

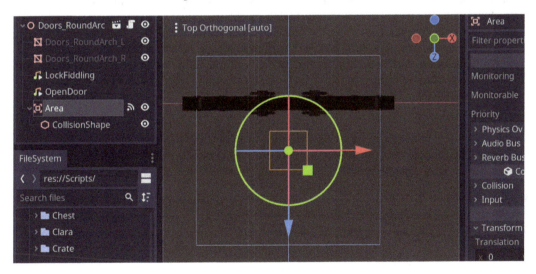

Figure 13.8 – This should be enough space in front of the door for Clara

The scene layout is pretty similar to the other examples you have created, but instead of one, there are two audio stream nodes. Their names indicate the kind of functionality we are trying to achieve. This time around, Clara standing in front of the door won't be enough by itself because we expect her to have found the key first.

Let's analyze the `Doors_RoundArch.gd` script and see how we are working it out. You can refer to this code block at `https://github.com/PacktPublishing/Game-Development-with-Blender-and-Godot/blob/main/Chapter%2013/Resources/Scripts/Doors_RoundArch.gd`.

We have a flag variable to keep track of whether the key has been collected. The value of this variable becomes true only when the `on_key_collected` function is run. All of this relies on whether the backpack variable emits the appropriate event, which is set up in the `_ready` function. That's why you are binding the backpack object to the door using the **Inspector** panel so that these two can communicate.

In the **body_entered** function, we check whether the intruding object is the player. This is where the flag variable comes into play. If the condition to open the door is satisfied, then we request the door opening sound. Otherwise, the game engine will play a sound file that indicates Clara fiddling with the lock.

> **One type of solution may not always cut it**
>
> The solutions we show you throughout this book may not always be ideal if your level or game structure is different. Even the game we are building right now might benefit from a drastically and much more efficient architecture. The concept of architecture means the hierarchy of game objects you lay out in your scenes, how scripts share common variables, and ultimately how your systems talk to each other. There is no golden solution, rather best practices that come with more exposure to coding, perusing forums, and attending conferences where seasoned developers share their battle scars.

We suggest you try both cases where Clara walks directly to the door to hear the no-go sound. Then, have her pick up the key, which is already notifying the player with its pickup sound. Lastly, she can go in front of the door again to hear the door creaking. That door sure needs some greasing!

Even though the squeaking sound makes us think the door is opening with some protest, we don't see it yet. So far, we've successfully mixed different disciplines we learned in the *Playing music and sound effects* and *Creating reaction spots* sections. It's time we added the missing animation component to our workflow.

Building simple animations in Godot

Back in *Chapter 5, Setting Up Animation and Rigging*, we discussed variances between Blender and Godot Engine for animation needs. In summary, we claimed that you'd be better off with Blender for animating anything more complex than bouncing balls and simple rotating objects. To drive the point home, we **rigged** and animated a snake model. Similarly, we have been using a humanoid character, Clara, done in Blender as well.

However, there comes a time when it might be suitable to animate some of the models in the game engine. The topic we have at hand is the opening animation of the arched door Clara is standing in front of. If you prefer so, you could still open the model in Blender, implement the necessary steps that represent the opening of the door, and reimport your work in Godot. It'll be no different than any other imported model that came with its animation.

For such a simple task, it's a bit of an overkill, though. We'll still use **AnimationPlayer**, but instead of triggering imported actions, we'll create our own by manually placing keyframes in the timeline to match the creaking sound we play when the door opens.

Creating the door animation

Before you start tackling any kind of manual animation in Godot, we suggest you take a closer look at the **MeshInstance** nodes the model uses. In our case, we are fortunate that there are only two. However, this might also be a problem too.

The model's mesh shows metal rings for grabbing and pulling to open such a heavy door. Sadly, they are part of the same **MeshInstance** nodes. This means that they can't be individually animated. To be able to do it, you'd have to go to Blender and separate those pieces and reexport the model. Then, you'll have more **MeshInstance** nodes you can work with. Keep in mind, though, that any one of these options is fine but comes with a trade-off. More individual objects often signal freedom, but they also clutter the **Scene** panel if you don't need them in the first place.

We're not concerned about the rings on the door for the time being. Our goal here is to learn the basics of animation in Godot, which starts by opening the `Doors_RoundArch.tscn` scene. After that, you will perform the following steps:

1. Place an **AnimationPlayer** node under the root. This will automatically bring up the **Animation** panel at the bottom. If not, press the **Animation** button in the bottom menu.

2. Press the **Animation** button in this panel's top area to bring a context menu and select **New** in the options. As a reminder, you used the **Load** option in that context menu in *Chapter 5, Setting Up Animation and Rigging*.

3. Type `Open` and press the **OK** button to confirm.

4. Set the animation length to `2.3` by typing it in the area between the clock and loop icons on the right side of the panel.

There are a lot of similar named buttons or options in the last set of steps. Thus, *Figure 13.9* will help you see what the editor will look like after your latest effort.

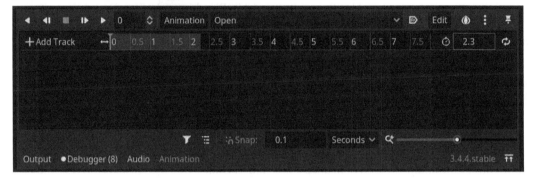

Figure 13.9 – Scaffolding for the open animation is done

The animation track is empty, but the groundwork is done. We need to tell **AnimationPlayer** how a specific property of an object is changing over time. To that end, you should do as follows:

1. Select the **Doors_RoundArch_L** node in the **Scene** panel.

2. Expand the **Transform** section in the **Inspector** panel. Press the key icon for the **Rotation Degrees** property. A confirmation popup will appear.

3. Press the **Create** button to accept the proposed changes.

4. Click and drag your mouse over the numbers in the timeline of the **Animation** panel. We want to set the time to the end of the animation, which is 2 . 3. Alternatively, you can type it in the area above the timeline to move the time marker.

5. Change the **Y** value in **Rotation Degrees** to - 6 0 and press the key icon again. There won't be a confirmation popup this time.

If you scrub the timeline back and forth as you did to move the time marker, you'll now see the door pivot around its hinges. Speaking of which, this was covered in the *Setting origin points* section of *Chapter 6, Exporting Blender Assets*.

Also, feel free to use the forward and backward play buttons to test the **Open** action. We'll trigger it programmatically soon, but we should take care of the other portion of the door first as follows:

1. Select the **Doors_RoundArch_R** node in the **Scene** panel.

2. Reset the time marker to 0 in the **Animation** panel.

3. Follow *steps 2–5* from the *preceding set of instructions* with only one difference. Mark the **Y** value as positive 6 0 this time since the directions are reversed.

After the two sets of changes, the editor will resemble what you see in *Figure 13.10*:

Figure 13.10 – Two sections of the door model have been keyframed, hence animated

This will add the necessary keyframes to the timeline at points where changes occur. Since we want the door to open in one go without any slowing down or stuck effect, we are not introducing more keyframes other than those we are using. If you fancy more complex scenarios, you can position the time marker along the track to where you want to introduce more keyframes.

The **Open** animation you have just created should run on a condition. We've already discussed and even implemented the necessary condition to a certain extent. However, we didn't really place the animation part in the door script. Let's do that right away.

Playing the door animation on a condition

Earlier in the *Interacting with the door* section, we attached a script to the door scene. This script had all of the necessary rules to check whether the player satisfied the conditions to open this door. We've also done a whole bunch of other things since then. So, let's summarize what we've got so far.

The arched door scene has an **Area** node that reacts to the player's presence. The door provides an auditory effect either way, but if Clara has already claimed the key, we expect the door to open with a creaking sound effect. Aptly named, we should trigger the **Open** animation. The change is simple enough, and it requires you to do as follows:

1. Open the `Doors_RoundArch.gd` script.
2. Replace `print("Open Sesame!")` with `$AnimationPlayer.play("Open")`.
3. Press *F5* to run the game. Have Clara first go for the key and then stand in front of the door.

Voila! A big obstacle in the way of going upstairs has been eliminated.

Although it's not possible to convey sound and visual effects via a still image, nevertheless, the following is the fruit of your hard work in *Figure 13.11*:

Figure 13.11 – Clara opened the door only after she collected the key from the backpack

If you move Clara away and come back near the door, the animation and sound will trigger over and over. Coming up with the necessary conditions to execute an event is important. However, it might sometimes be equally important to stop it from happening again. You might have already noticed a similar, and maybe annoying, repeating behavior with the candles as well. Some effects should only fire once.

We still have quite a few things to do in this chapter. That's why we will give you a quick guideline for eliminating this kind of repeating behavior. By nesting or combining `if` blocks, not only can you make sure the condition has been met just then, but also that it has been met before. For this, you might want to take advantage of simple Boolean variables. If the solution doesn't come to you, you can always check the GitHub repository for the finished work.

What's left for Clara to do at this point? Well, she's currently standing there waiting to go upstairs. In this context, upstairs means loading another level, which we will discover in the *Loading another level* section later. For the time being, we still don't know exactly when we are supposed to load the next level. Let's see how we can determine that.

Waiting for the door animation to trigger an event

It's tempting to load the next level when we start opening the door. That being said, you've worked hard to keep track of what Clara has been doing as a precondition to start the door's opening animation. If you switch to a new level right away, the animation will be for naught.

Instead, we should wait for the **Open** animation to finish. Only after that does it make more sense to switch things up. There are two common but equally awkward ways to do this. We'll discuss both, so you get to know them before we dismiss them for the sake of a better alternative, and they are as follows:

- `yield`: You can add `yield($AnimationPlayer, "animation_finished")` after you trigger the **Open** animation. Whatever comes after the `yield` line, such as loading a new level, will have to wait for the animation to be finished. This is, in a way, like holding the line. Nothing else will happen unless, well, the program yields. This concept will change in Godot 4 in favor of the **await** command, which is a more permissive architectural choice than blocking things during the execution of your code.

- **Timer**: An alternative to `yield` where you are still letting things run is introducing a **Timer** node to your **Scene** tree. This is just like any other node you could add. Its **Wait Time** field in the **Inspector** panel could be set to when you want it to go off, in our case, 2.3 seconds, since that's the length of our **Open** animation. Then, once the time is out, this node will fire a **timeout** signal for which you can write a listener.

  This method's usage in our situation would be to start the timer as soon as you initiate the **Open** animation. Since the timer's **Wait Time** would be synced with the action you are playing, it would look like loading a new level right after the action is finished.

We will not use either of these methods because why would you make your life more complicated when there is already a way to accomplish something with the toolset you are familiar with? Instead of switching gears, we'll see how **AnimationPlayer** can still help us as follows:

1. Add the following function somewhere in the `Doors_RoundArch.gd` script:

   ```
   func load_level():
       print("What level?")
   ```

2. Select the **AnimationPlayer** node and expand a context menu by pressing the **Add Track** button.

3. Choose **Call Method Track** among the options. You'll be presented with a list of nodes to pick from. So, select the root node, **Doors_RoundArch**, on the upcoming screen.

4. Move the timeline marker to 2.3 seconds. Right-click where the blue timeline marker meets **Functions** for the **Doors_RoundArch** entry in the animation tracks. To get a better idea, refer to *Figure 13.12* to see the location we are talking about.

5. Search and choose **load_level** from the upcoming list. Press *F5* to run the game and follow the necessary steps as before to open the door.

Everything will be exactly the same, except when the door animation is finished playing the **Open** sequence, the `load_level` function will run too. Since showing the door animation won't make sense, we'd rather show you the editor's status as mentioned in the fourth step:

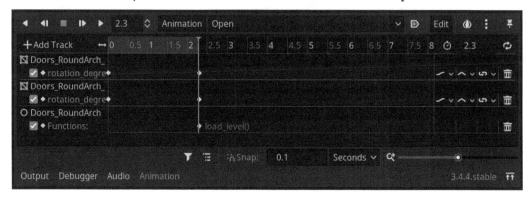

Figure 13.12 – The load_level function will be triggered when the timeline arrives at the keyframe we set

The last frame of the **Open** action is where we are firing the function responsible for loading the next level. For now, it's printing only a statement. We'll be looking into swapping our current level with a new one later in the *Loading another level* section.

While we are still working on building simple animations, we could take care of the light sources that kind of look static.

Let there be flickering lights

The work we did with the sconces and candles for introducing the **Light** nodes to our game in *Chapter 10* , *Making Things Look Better with Lights and Shadows*, didn't include animations. Nevertheless, we've been gradually improving everything else ever since.

Consequently, it would be nice to add some oomph to our light sources as follows:

1. Open `Sconce.tscn` and add an **AnimationPlayer** node to the root.

2. Introduce a new action. Choose `Flicker` for its name.

3. Set the length to 2 seconds. Also, turn on **Animation Looping** and **Autoplay on Load**.

4. Press the **Add Track** button and choose **Property Track**. Select **OmniLight** from the list that pops up. This will display another list to pick from.

5. Pick **omni_range**. Right-click the track in the **Animation** panel at 0.0, 0.4, 1.3, and 1.9 seconds to open a context menu and select **Insert Key**.

6. Select each one of these keyframes and enter 8, 6, 7, and 5, respectively, in their **Value** property in the **Inspector** panel.

7. Press *F5* and have Clara light the sconces. They should start to flicker.

Before we discuss a more refined and advanced version of what we have done, the following is what we have in the **Animation** panel:

Figure 13.13 – The Flicker action has been defined for OmniLight in sconces

Things now must look more organic when you light the first sconce. Then, perhaps after the second or the third one, the cozy flickering effect will look disturbingly repetitive, won't it? If only there was a delay between different sconces so they wouldn't all fire the **Flicker** action at the same time.

Achieving that will be relatively easy, but we suggest you first copy the **AnimationPlayer** node inside Sconce.tscn and paste it into both the Candles_01.tscn and Candles_02.tscn scenes. It'll be easier to notice the effect of randomness when we use the animation everywhere.

When all of the light sources are lit, the whole level will look like it's pulsing. Let's see how we can break the unanimity and introduce some randomness to what we have, as follows:

1. Turn off **AutoPlay on Load** in **AnimationPlayer** for all of the three scenes you are using it for.

2. Open the LightSwitch.gd script and alter the _process function as follows:

```
func _process(_delta: float) -> void:
    $OmniLight.visible = is_lit

    if is_lit:
        yield(get_tree().create_timer(randf()*2.0),
            "timeout")
        $AnimationPlayer.play("Flicker")
```

All our light sources share this script. So, the changes will apply to all instances. While we were not in favor of using the `yield` command, it was relatively harmless to do so in this case. The last three lines tell the engine to create **Timer** on the fly and it randomly picks **Wait Time** for it between 0 and 2 seconds. When this timer goes off, the **Flicker** action plays.

Although you copied and pasted the same **AnimationPlayer** node that forced the light sources to share the same length and keyframes with exactly the same values, since the **Flicker** action for each light starts with a delay thanks to our latest change, it will induce enough visual differences.

Additionally, if you want to be really fancy, you could add another track such as **light_energy** to vary the brightness of the light sources.

Wrapping up

Slowly but surely, you will have a more complete and believable feeling game by introducing small variations here and there, either by placing them in the world in a non-repeating pattern or by animating some of the game objects' key features.

Sometimes the method to do this will be completely different. For example, the shader we are using to simulate the body of water doesn't use a node such as **AnimationPlayer**, but we still have motion. That being said, it's disillusive to have that boat look so still while the water is in motion. With the knowledge you have gained in this section, we suggest you turn the boat model into a scene and animate it to show an oscillating motion like a boat would do.

While you should feel confident that you know how to animate the basic properties of game objects, you have left out something important: Clara was supposed to head upstairs. Let's help her do that.

Loading another level

Before we started to animate the light sources in the *Let there be flickering lights* section, we were ready to move Clara upstairs. To that end, we used a nifty feature of the **AnimationPlayer** node to fire the `load_level` function, which printed a statement to the **Output** panel, a substitution for the real thing. In this section, we'll investigate how to swap the existing level with another.

Let us remind you that our current level, `Level-01.tscn`, is instanced inside the `Game.tscn` scene, which is holding a **Camera** and an **AudioStreamPlayer** type of nodes. Godot has a built-in function, `change_scene`, that can change the current scene to another scene. However, this might be dangerous since it'll replace the entire structure. In our case, this is not `Level-01.tscn` but everything in `Game.tscn` because that's the main scene.

The solution we'll offer is a process that's operational at a higher level than `Level-01.tscn` itself. Ideally, your scenes should notify a higher authority of the changes they would like to introduce to the overall system. As it happens, this could very well be the `Game.tscn` scene via which not only can you use it to load a new level, but you could also be taking care of other stuff in your game such as keeping a log file, contacting a database to store important changes, or even reaching to a third-party service to show ads.

Now that we've established the importance of the Game.tscn taking over the task of loading a new level, how are we going to let it know when to do it? You have used signals before to facilitate a way between different game objects to know each other. This involved placing a reference of an object inside another by exposing a script variable to the **Inspector** panel. Although we could still try this, there is a better way.

Using an event bus

When we expose variables to the **Inspector** panel so that scripts can recognize other game objects to be able to connect to their signals, we are coupling things, in a sense. When the number of objects and signals grows, this method will be difficult to maintain. There is an alternative, a concept called **event bus**, that might be helpful in an ever-growing list of dependencies.

We'll revisit this concept in more detail in the *Further reading* section since the notion is part of a much bigger family of options available to you. For the time being, we'll be satisfied with a practical application of it. This is what it entails:

1. Create an EventBus.gd script in the Scripts folder. Add the following line to it:

    ```
    signal change_level(level)
    ```

2. Open **Project Settings** and switch to the **AutoLoad** tab.

3. Use the button with the folder icon to find the EventBus.gd script.

4. Press the **Add** button to add this script to the list underneath.

Figure 13.14 shows what the editor will look like.

Figure 13.14 – Our first singleton is set up and ready to use

We have just added a script to the **AutoLoad** list. A **singleton** is also another common name that is used in the industry for this concept. It means that there can only be one instance of the script. Besides the conventional description, in a Godot-specific context, as soon as you introduce it to the **AutoLoad** tab, there will always be one and only copy of this script; it will also be loaded for you and be made available to all of the constructs in your project.

So, who's going to make use of this new script since it doesn't seem to be attached to anything? After all, it just exists there, but since **AutoLoad** makes it accessible at all times, we can use it when the door animation is finished.

Let's reassess our work from the *Waiting for the door animation to trigger an event* section. When we run and wait for the **Open** action in the Doors_RoundArch.tscn scene, **AnimationPlayer** eventually triggers the load_level function. There is currently a line of placeholder code in the body of that function in the form of printing a short statement: **What level?**

That's where we originally intended to load the next level. However, in light of the discussion we had in the opening lines of the *Loading another level* section, we now want to delegate this to the Game.tscn scene. To that end, we have created an EventBus.gd script that will communicate our request to the relevant recipient. Therefore, you will have to make the following change:

1. Open the Doors_RoundArch.tscn scene.

2. Update the load_level function as follows:

    ```
    func load_level():
        EventBus.emit_signal("change_level",
                             "Level-02.tscn")
    ```

In our earlier efforts, game objects were directly using the emit_signal command. For example, the backpack was emitting a key_collected signal. Here, we generalize the idea. We no longer care about knowing which object is emitting. We use a high-level construct such as EventBus to do this for us. *Figure 13.15* shows a diagram of the new architecture we are proposing.

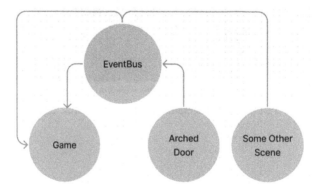

Figure 13.15 – We no longer need to couple structures anymore thanks to EventBus

In the backpack example, the emitted signal was directly captured by the door so that the game could decide whether the player has completed a necessary condition. So, similar to how communication works in real life, there are two main parts to an event: an emitter and a receiver. We've made updates to the emitting situation. Let's see what we can improve at the receiver's end.

Listening to the EventBus signal

Going back to the relationship the door and the backpack objects had, the backpack wasn't aware of the door, but the door had a field we set in the **Inspector** field to reference the backpack. So, when the backpack emitted an event, the door was already keeping an eye on the backpack in a manner.

We are now trying to stay away from this type of architecture. Instead of directly using an object to emit an event, we tell the EventBus to do it for us. However, who is the door in our new example? In other words, who is listening to our event and how? The short answer is the Game.tscn scene.

Let's implement some code first. Sometimes, it serves the purpose of showing instead of telling. Then, we'll explain the rationale behind it. The following steps show what you should do after you open Game.tscn:

1. Create a new **Spatial** node under the root node. Rename it as Level.

2. Drag the **Level-01** node into this new **Level** node.

3. Make a new script as Game.gd and attach it to the root node. You can save it alongside the scene file. Then, you type in the following code:

```
extends Node

func _ready():
    EventBus.connect("change_level", self,
                      "change_level")

func change_level(level:String):
    var new_level = load("res://Scenes/" +
                      level).instance()

    $Level.remove_child($Level.get_child(0))
    $Level.add_child(new_level)
```

Do you see that _ready function where we make use of the EventBus architecture? That's the sweet part. This way, neither Game.tscn nor Doors_RoundArch.tscn need to know anything about each other. They share and deal with their responsibilities through EventBus.

Somewhere, at some point, a structure may fire a `change_level` signal. That is all we care for, and after we express our interest in it, we also prepare ourselves for what to do with it, in case the event comes to fruition. If that's the case, we handle it inside the `change_level` function.

> **Naming conventions**
>
> Some people keep their signal and event handler (function) names the same for the sake of treating the function as an extension of the signal. Godot's signal bindings will add an `_on_` prefix, though. Keeping your own event handlers' names the same as the signal name might help you distinguish them from Godot's own bindings. However, you could always follow Godot's naming convention in your bindings too.

Let's now analyze what's going on in the `change_level` event handler. When we fired the signal in the arched door scene, `EventBus` was passed a parameter in the form of a string: `Level-02.tscn`. The first line in the `change_level` function looks up and loads this string in the project's `Scenes` folder. After finding a match and creating an instance of it, we want to store this new scene because we still have some work to do with the current scene. We should dispose of it before we add the new scene.

Since we've made some changes to the **Scene** tree, the current level, **Level-01**, is now inside a **Level** node that acts like a receptacle. Thus, we are instructing it to first find then remove its only child with `$Level.remove_child($Level.get_child(0))`. Only after that do we add the new level.

There is only one thing left for you to do. Press *F5* and have Clara go through all of the steps necessary to trigger the door's opening. As soon as the door is open, the game will take you upstairs to a new level. You should expect to see what *Figure 13.16* shows.

Figure 13.16 – Welcome to our new level

Congratulations! You have guided Clara to find her way in the darkness to collect a key that unlocked the door to this new level. She can continue her adventures from here. Is that a chest over there? There is a trapdoor right in front of it though, so watch out for that. Using the tools that we have shown you, you can go on and create new conditions and obstacles for the player to tackle. It's up to your imagination.

We'll now dedicate the rest of this chapter to discussing some of the choices you've made by following our guidelines and what you could also do differently.

Discussing some of the choices we can all make

Our goal in this book is to teach you just the necessary parts of Godot Engine to build a simple point-and-click adventure game. It's a simple statement, and yet it entails two separate efforts. On one hand, we should teach you as much as possible about the game engine without making it look like you are reading documentation.

On the other hand, the game we planned to build must be advanced enough but also simple to the point that you can easily follow its progress by reading as little as possible. Also, the fact is that there are only so many pages in a book. Thus, some of the choices we made during the production of the game were limited by these factors.

You might also face similar but different limitations and conundrums in your own projects. An early plan, even the worst one, might often be better than not having a plan at all. Even then, some cases might be really hard to nail and prepare beforehand, such as making your gameplay fun or achieving a decent user experience.

For example, the level switch is technically done. However, the change is happening so abruptly that the player might want to feel a moment of respite to collect their thoughts and savor their journey throughout the level. You can easily achieve this by extending the animation length and pushing the `load_level` function to later frames. It might look like there is a healthy pause between the door animation and the loading of the next level.

Even better, having the screen fade out before the switch actually happens might be a good idea. In fact, this might even be useful from a technical point of view. Our second level is so small, thus it's easy to load it from the disk. However, in more ambitious projects, your levels might be chuck-full of game objects waiting to be loaded.

Furthermore, if your game loads previous sessions, you will have to reset your game objects' states to their last known values. A generic loading screen in between switching levels or loading a previous game session might be a much better architecture. By following this practice, you'll most likely find yourself abstracting more and more systems from more directly implemented systems.

Thus, this is perhaps the most valuable piece of advice we can offer you: if you are feeling stuck or unsure of how to tackle a topic, first focus on the special case and its implementation, then try to generalize it if possible and necessary.

Summary

This was another chapter with a lot of moving parts that incorporated so many different aspects of the game engine. Let's break down some of your activities that helped to add the finishing touches on so many things we carried over from the previous chapters.

First, you tackled background music and sound effects. You had already seen the usage of sound in *Chapter 8, Adding Sound Assets*, which covered simple scenarios. In this chapter, you've learned how to use sound assets in a proper context.

Next, you reexamined a topic you saw in *Chapter 12, Interacting with the World through Camera and Character Controllers* – player detection. This time, you used **Area** nodes as trigger zones since there would not be direct player interaction, such as mouse clicks and motion. Instead, Clara triggers predetermined events when she's in the right zone.

You were also able to communicate information between game objects, essentially separate and distant systems, when an **Area** node was actively used. For instance, when the player reached the backpack, the condition to open the door was satisfied. The backpack let the door know what was going on through the use of a custom signal.

You symbolized the pickup of the key with a sound effect. Perhaps, a short piece of animation would have been used to display a 3D key moving up and fading out. Sometimes, an icon appears at the bottom of your monitor and finds its place in what's called a **quickbar** in some games. Both approaches are fine, but we didn't want to do either one of them.

Since this chapter was supposed to teach the creation of animations in Godot, we wanted to show off cases that were sufficiently complex, such as flickering light sources or opening two sections of an arched door, rather than simply moving a key up in the game world. We believe our effort has a more didactic value that you can transfer to other simple use cases.

After finishing simple animations, particularly the door's opening action, it was time for Clara to go upstairs. To achieve that, you looked into swapping the current level with a new one. Although you could have achieved this by letting game objects pass information between each other, you were introduced to a more generic way of doing this via an `EventBus` architecture.

Even though there is still one more chapter, this is the moment you should pat yourself on the back. You have built a fully functional, however small, point-and-click adventure game. The following chapter will show you how to export your game. We'll also discuss what other options you can consider on your game development journey.

Further reading

As promised, we want to share with you a few words on the artistic aspects of sound management. Sometimes, a piece of music will have a high tempo. It means it'll have a higher value of **beats per minute (BPM)**. Depending on the game or the level you are building, you might want to select or create your music with the most appropriate BPM value to convey the best emotions.

There are also situations where gameplay will ask for a mix between a higher and lower tempo. This is common in role-playing or action games where players would like to feel they are under tension when they get involved in a sticky situation. For example, it would absolutely break the immersion if your burly, gun-toting player character is hiding behind a cover under heavy enemy fire when classic or chillout music is playing in the background. Likewise, when all is supposed to look calm between two action zones, if the game is playing a piece of high-tempo music, you will needlessly stress out and confuse your players.

Luckily, there are plenty of courses on this topic on Udemy. Giving a list of courses here would do injustice to all of the others we couldn't mention since the list is long. We suggest you look it up on their website by using the **music for games** keywords.

Last in the sound management topic is the use of supplemental technologies. Either of the following two will help you create on-the-fly solutions to ever-changing circumstances if your game can't make use of prearranged sound assets:

- FMOD
- Wwise

We also briefly mentioned artificial intelligence in this chapter. This is a vast topic, but a pertinent list of books would be the following:

- *AI for Games* by Ian Millington
- *Behavioral Mathematics for Game AI* by Dave Mark
- The *Game AI Pro 360* series by Steve Rabin:

  - *Game AI Pro 360: Guide to Character Behavior*
  - *Game AI Pro 360: Guide to Movement and Pathfinding*
  - *Game AI Pro 360: Guide to Architecture*
  - *Game AI Pro 360: Guide to Tactics and Strategy*

The `EventBus` solution we presented in this chapter is frequently utilized in many programming circles. It's sometimes called a **Publish/Subscribe** model or an **Observer** pattern. Referring to *Figure 13.15*, imagine you replaced `EventBus` with the post office. When a magazine you are subscribed to has its latest issue coming out, the publisher will notify the post office and you'll be delivered your subscription.

Since the inception of computer science, and more particularly software programming, developers have noticed problems that exhibited a particular behavior or nature. Solutions to these common problems are called **design patterns**. There are a lot of resources out there that deal with this topic in the framework of classic software. However, game developers have also gotten some love in recent years. Regardless of domain specificity, a few examples are the following:

- `https://gameprogrammingpatterns.com`
- `https://www.udemy.com/course/design-patterns-for-game-programming/`
- *Head First Design Patterns: Building Extensible and Maintainable Object-Oriented Software* by Eric Freeman
- *Learn Design Patterns with Game Programming* by Philippe-Henri Gosselin

14
Conclusion

Congratulations!

You have built a point-and-click adventure game that utilizes 3D assets, incorporates camera and character controllers that respond to player inputs, triggers visual and sound effects for feedback, follows player progress, and loads a new level.

This chapter will cover a topic that is usually covered when you come to the finish line. We will show you how to export your game so that you can share it with the rest of the world. That being said, we'll also discuss reasons why you might want to export more frequently than just waiting until the end.

After that, we'll be fully done with the technical parts of the engine. Hence, we'll present a few pieces of advice, more like guidelines you can follow in your development cycle to be efficient either before you start your projects or during them.

Lastly, you will look at a few game genres for which you can use Godot Engine. Every game engine is usually built around at least one strong and a few core needs. That being said, most engines worth their salt also support the most expected features. You'll see how some of the knowledge you have gained throughout the book could be expanded upon in new areas.

This is going to be a relatively short and, most definitely, less technical chapter. Nevertheless, we still have the following topics to tackle:

- Exporting your game
- Offering different gameplay experiences
- Discovering different genres

By the end of this chapter, you'll have learned how to export your creation, evaluate different options you can offer to your players, and – finally – find a list of genres you can consider using Godot Engine for.

Technical requirements

There won't be any new resources in this chapter. If you prefer, you can continue your own work from the previous chapter or peruse the content we keep in this book's repository at `https://github.com/PacktPublishing/Game-Development-with-Blender-and-Godot`.

Exporting your game

So, you have a game. What now? You can keep running the game in the editor, as you've been doing all along. At some point, though, you'll most likely want to show it to your friends and family or even deploy it somewhere public for everybody to look at it. This section will teach you how to export your game so that you can share your creation with the rest of the world.

Although we'll only cover how to do it for Windows, Godot Engine is also capable of exporting your game to the following platforms:

- Android
- iOS
- HTML
- Linux
- macOS
- **Universal Windows Platform (UWP)**

Although exporting is usually a simple process, it would be wise to check the documentation since updates that platforms receive sometimes change the steps you must take. You can find the most comprehensive list of instructions here: `https://docs.godotengine.org/en/3.4/tutorials/export/`.

> **What about consoles?**
>
> Consoles are not part of the aforementioned list because they lie in a somewhat gray area due to licensing. As a developer, you need to be in touch with a console producer and sign agreements to have access to their tools and kits. In essence, although there is still some technical aspect to this, it also has some moving parts in the legal department.

Before we start tackling Windows-specific export settings, we need to add or change a few things in our project.

Preparing your project for export

By default, Godot doesn't launch your games in **Fullscreen** mode even though it's something most games use. While in the end, we will make our game cover the whole screen, it's worth discussing a few other options you will see when you open **Project Settings**. More specifically, you'll see two features when you visit the **Window** section under the **Display** group, as follows:

- **Resizable**: This option makes your game screen resizable, just as you would be able to resize any other application that's not in **Fullscreen** mode. This is on right off the bat, so turn it off.

- **Borderless**: When your game is not running in **Fullscreen** mode, it will have to have borders defined by your operating system. Having this option on will remove those borders and the header of the window. By the way, most modern desktop applications—such as Slack, Discord, and likewise—use this feature these days.

We suggest you turn on the **Fullscreen** option and the other two that we just talked about off. After that, this is what our **Project Settings** screen looks like:

Figure 14.1 – The project settings we are using before we export our game

So far, we've focused only on building the game itself without worrying about the intro, game settings, or credits screens. These can be constructed just like any other Godot scene. Then, once you figure out the flow between these scenes, you can use the `change_scene` function to transition to the one the player is asking for. Alternatively, you can keep some of these screens as hidden scenes inside the `Game.tscn` file and turn their visibility on as requested.

Since our game will now run in **Fullscreen** mode, you won't be able to terminate it by using the operating system's buttons. In Windows, pressing the *Alt + F4* key combination will exit the window. We need to provide a far better way for the player to quit the game.

Creating a mechanism for turning the game off

Movies end with the production companies' logos and actors' names on a theater screen. Unless you are really intent on looking at the credits, you will consider this moment as your cue to get up and leave the theater. Either this way or if you want to terminate your movie experience early on at any moment you want, you have the freedom to leave the premises.

A similar situation would happen with the click of a button if you were consuming a movie with a video player on your computer. When we run our little game in **Fullscreen** mode, since there won't be any button around to click, this is something you have to present to your players in different forms.

This is usually done by pressing *Esc* on the keyboard to reveal a screen—sometimes blocking the game screen and sometimes as an overlay—so that the player can either go into the game's settings or load a different game session and obviously quit the game.

We will implement only the *Esc* press part in this section and treat it as the player's desire to quit. To that end, we suggest you open the Game.gd script and add the following lines of code to it:

```
func _input(event):
    if event.is_action_pressed("ui_cancel"):
        get_tree().quit()
```

You might have been expecting to see *Esc* in that if block. It's there but as an **identifier**. If you go to **Project Settings** and bring up the **Input Map** tab, you will see a list of shortcuts that are mapped to easily comprehensible names. The following screenshot shows a portion of **Input Map**:

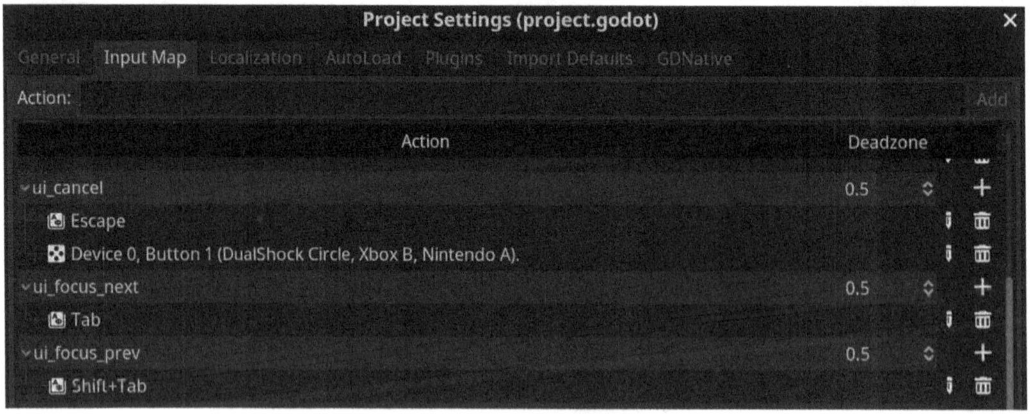

Figure 14.2 – The Input Map tab is part of Project Settings

If you're building games that allow your players to use multiple input devices, then configuring **Input Map** will be tremendously helpful. For example, you could set it so that a game controller or a joystick's button press means the same thing if the player wishes to exert the same behavior with a keyboard. It's a neat way of consolidating different inputs under one name you can easily follow in your code.

We took care of screen sizes and letting the player quit the game, so we should be all set for exporting our game.

Configuring Windows export settings

Godot's download size is extremely small compared to other game engines. One of the reasons for this is that it doesn't come loaded with export packages. Platform requirements sometimes change and Godot's specific functionalities must conform to their guidelines, so it makes sense to download and get updates on export packages as you go.

Since we've never exported a game, there is no export package in our setup. To get one, press the **Editor** button in the top menu to access the **Manage Export Templates** setting. When you bring it up, you will see an interface with which you can download and install the right package for the version you are using. The following screenshot shows the current state of export templates:

Figure 14.3 – This screen will help us download export templates

You should press the **Download and Install** button and wait. Once that's done, you could press the **Close** button in that interface. Next in our export efforts is to work with the **Export** settings, so follow these steps:

1. Press the **Project** button in the top menu and select **Export** among the options.

2. Press the **Add** button and choose **Windows Desktop** among the options.

3. Fill out the **Export Path** option by using the button with the folder icon. We chose to export it to a `Build` folder outside the project files, so we defined it as `../Build/Clara.exe`.

4. Press the **Export Project** button in the bottom part of the **Export** interface.

5. Turn off the **Export With Debug** option near the bottom. Confirm your file path and press **Save**.

Before we move on to explaining things, here is a visual representation of some of the steps you had to take for exporting:

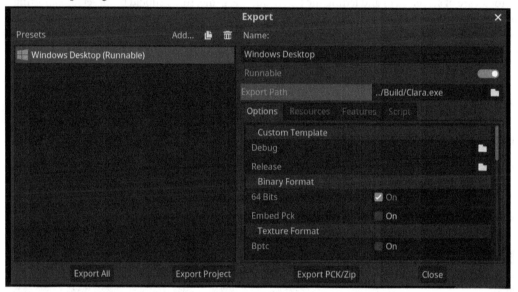

Figure 14.4 – Some of the export settings for Windows

These steps, assuming your Windows is running on a 64-bit machine, will export your game to the folder you defined. When you run the executable, you should be playing the game just as you were while you were developing it in Godot. Pressing *Esc* will terminate the program and take you back to the operating system.

You might have noticed an extra file with a PCK extension besides Clara.exe. If you want to keep those two together, you can turn on the **Embed PCK** option in the **Export** settings, yet keeping things separate might be a good idea too. Godot keeps your game's resources in a separate package file and uses it when you run the executable.

Why or when would this be useful? If you want to enhance your game with more content, you can create only content packages and instruct the game executable to pull them in. Your next DLC might be just around the corner, and this is a useful mechanism toward that goal.

Not only do you have a finished game, but you can also ship it! Exciting, indeed. Although we have provided mostly technical instructions, we feel it would also be valuable to share a few words about different gameplay experiences you can offer to your players.

Offering different gameplay experiences

Sometimes, it's OK to use prototype assets or another artist's creations so that you can focus on fun. We are saying this with a word of caution because we'll always advise you to be sure of the license of the assets you are using. That being said, the topic we want to discuss is what you do with the assets once you have access to them.

The `Models` folder contains extra assets that we didn't use throughout the book. When you were constructing the first level in *Chapter 9, Designing the Level*, we mentioned that you could use some of those other assets. Maybe you did and had to follow the instructions in later chapters based on your own conditions, especially with input detection, pathfinding, and likewise.

At some point, as with right now in the last pages of this book, you might find yourself at a loss for coming up with what more to add to your game.

Having an iterative creation process

Some people find it much more empowering to have visual assets laid right in front of them. The creative juices start flowing when they look at different objects' size-and-shape relationships. Then, there are others who find this inconvenient and getting in the way of drawing out a proper plan. If they figure out what needs to be done, they can start altering assets or looking for new ones. Finally, a mix of both of these approaches might work.

In the end—specifically, if you want to go commercial with your work—you've got to keep the player in the center of your workflow. Quick iterations followed by early and frequent playtesting might be what you need. The ramifications of some of your choices mixed with players' expectations from the game might create a lot of stress, so be aware of this. We'll give you an example by using the assets and layout of the second level.

There are currently two bookcases on that level: an upright one and a knocked-down one. This is a relatively cheap and effective storytelling method. Why is one bookcase on the floor? Perhaps there was a calamity, but we don't know. Is it going to be moved out of the way? If you, as a developer, want it or the playtesting shows it's a strong request, then you have to spend more time in Blender or Godot to come up with an animation plan for the bookcase. Clara will most likely need another action that shows her lifting up and moving the bookcase. If she shouldn't because it's unlikely that she can lift up such a heavy object, then you either need a tool or a companion that can help her.

One simple change or request, and you will be inundated with a series of tasks. Unfortunately, not all these changes will be visual either. You'll have to account for the programming parts where you have to keep the state of the bookcase still on the floor or moved out of the way.

Ultimately, as the creator, you've got to ask yourself where this effort might be leading. If you could take this idea to have Clara access another level or a secret used in the game—in other words, mix it with something that already exists as a mechanic—you can replicate it with the minimum number of steps; it might be worth it.

So, it's always a trade-off. As much as you should honor fun and your players' requests, you should approach it carefully and also consider what works best for you.

As we are wrapping up our book, let us discuss which other things you can do with Godot.

Discovering different genres

Even though Godot Engine is known for creating quality 2D games and other well-known engines are preferred for building 3D games, you have seen that Godot is actually quite capable of building a 3D game. This is going to change for the better when Godot 4 comes out.

Until then, what else can you do with Godot? You can build any kind of game with it, to be honest. There has also been a recent trend to build desktop applications using Godot Engine. However, we will consider these cases as extraordinary and instead focus on some more commonly known genres that employ 3D features, as follows:

- Simulation and strategy games: When you used raycasting to detect user input, it was done so that Clara could move to a particular spot with pathfinding. In a simulation or strategy game, either on a grid or free-move structure, your selected unit or units could move to their designated destination in a similar way. You could even combine a turn-based feature on top of this where you keep track of which side's units have already moved.

- Racing games: Godot already has a **VehicleBody** node to simulate the behavior of a car. Isn't that nice! By appropriately placing a **Camera** node inside a **MeshInstance** node and combining the mechanics of a **VehicleBody** node, you could be building the next awesome racing game. Start your engine, Godot Engine, and vroom!

- First-person shooters: A classic example that could definitely be built with Godot Engine. You'll be using raycasting a lot in this type of game where you detect whether bullets connect with objects. If they do, maybe a good mix of technical and creative problems lies ahead of you. Should bullets penetrate or destroy every object the same way?

- Role-playing games: This is similar to First-person shooters, so it could be done. In this genre, you generally have a lengthy narrative to present to your player. Also, you've got to keep track of where the player is in the story and whether they have met some of the conditions to reveal the following parts of the story or the outcome of a puzzle. We haven't discovered this in this book, but it would be wise to check out `Resource` as a useful Godot mechanism to facilitate content-heavy games.

- Multiplayer/Co-op: This is not a genre by itself, since any genre can be made multiplayer or co-op. However, there are some games where the experience won't be the same without networking, so we had to mention this separately. Godot has networking parts you can use to connect to third-party services or have two computers in the same network connect to each other.

These are some of the genres that can most definitely be made with Godot. You can also include some other genres such as puzzles or sports games, or any other subgenre that uses 3D assets.

Summary

As we are concluding our book in this chapter, your game project is also coming to an end. Hence, we opened it by showing you the necessary steps for exporting your game. Even though it might seem like you'd tackle this phase once your game is built, as was mentioned in the *Iterative creation process* section, it might be wise to export your game often and share it with others for frequent feedback.

The rest of the chapter was dedicated to discussing different approaches you can take in your game development efforts, best practices, general guidelines, and—finally—getting to know different genres you can target.

You've come a long way in your game development journey. It started with Blender in the first five chapters and continued with a few transitional chapters until you fully switched to building a game with Godot Engine. Hopefully, you now have a much better opinion about how things work in both applications. Also, if you have some prior experience, we hope that this book has increased your confidence level in some areas.

As we are leaving you, we wish you the very best in your future efforts, and may your code compile the first time!

Further reading

You might have noticed that the exported game is using Godot's icon. It would be nice to have your own custom icon. There are several moving parts to this, but it's possible. The instructions are listed at `https://docs.godotengine.org/en/3.4/tutorials/export/changing_ application_icon_for_windows.html`.

If you would like to deploy your game for feedback purposes instead of sending files over emails or chat applications, you can use the following platforms:

- `https://itch.io`
- `https://gotm.io`

The latter URL is especially useful in our situation because that platform also hosts Godot game jams. For PC games, Steam is a big marketplace, but the aforementioned places might work faster than signing up and going through the application process on Steam.

Index

Symbols

3D model
  about 4
  parts 5
9-slice scaling
  about 205
  reference link 205

A

Action Editor 97
actions
  blending 254
  playing, for player 253, 254
Adjustments feature
  properties 195
Ambient Occlusion (AO)
  about 72-74
  properties 193
AnimationPlayer component 79
AnimationPlayer node 130
animations
  actions, separating 131-133
  AnimationPlayer node 130
  blending 254
  building 78

building, in Blender 79
building, in Godot 274
building, in Godot Engine 79
creating 83
creating, for Godot 98
importing 128, 129
MeshInstance node 130
Skeleton node 130
triggering 249-251
Area light
  about 66
  properties 69
armature 83
audio
  gameplay experience, increasing 145
  playing, in Godot 142
audio bus 261
audio file converters
  references 138, 139
AudioStreamPlayer 142, 259
AudioStreamPlayer2D 142, 259
AudioStreamPlayer3D 142, 258
AutoLoad 260

Hi!

I am Kumsal Obuz, the author of Game Development with Blender and Godot. I really hope you enjoyed reading this book and found it useful for increasing your productivity and efficiency in Blender and Godot.

It would really help me (and other potential readers!) if you could leave a review on Amazon sharing your thoughts on Game Development with Blender and Godot.

Go to the link below or scan the QR code to leave your review:

`https://packt.link/r/1801816026`

Your review will help me to understand what's worked well in this book, and what could be improved upon for future editions, so it really is appreciated.

Best Wishes,

Kumsal Obuz

`Packt.com`

Subscribe to our online digital library for full access to over 7,000 books and videos, as well as industry leading tools to help you plan your personal development and advance your career. For more information, please visit our website.

Why subscribe?

- Spend less time learning and more time coding with practical eBooks and Videos from over 4,000 industry professionals

- Improve your learning with Skill Plans built especially for you

- Get a free eBook or video every month

- Fully searchable for easy access to vital information

- Copy and paste, print, and bookmark content

Did you know that Packt offers eBook versions of every book published, with PDF and ePub files available? You can upgrade to the eBook version at `packt.com` and as a print book customer, you are entitled to a discount on the eBook copy. Get in touch with us at `customercare@packtpub.com` for more details.

At `www.packt.com`, you can also read a collection of free technical articles, sign up for a range of free newsletters, and receive exclusive discounts and offers on Packt books and eBooks.

Other Books You May Enjoy

If you enjoyed this book, you may be interested in these other books by Packt:

Godot Engine Game Development Projects

Chris Bradfield

ISBN: 978-1-78883-150-5

- Get started with the Godot game engine and editor
- Organize a game project
- Import graphical and audio assets
- Use Godot's node and scene system to design robust, reusable game objects
- Write code in GDScript to capture input and build complex behaviors
- Implement user interfaces to display information
- Create visual effects to spice up your game
- Learn techniques that you can apply to your own game projects

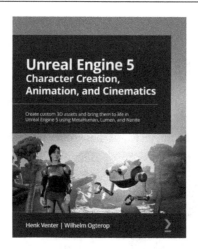

Unreal Engine 5 Character Creation, Animation, and Cinematics

Henk Venter, Wilhelm Ogterop

ISBN: 978-1-80181-244-3

- Create, customize, and use a MetaHuman in a cinematic scene in UE5
- Model and texture custom 3D assets for your movie using Blender and Quixel Mixer
- Use Nanite with Quixel Megascans assets to build 3D movie sets
- Rig and animate characters and 3D assets inside UE5 using Control Rig tools
- Combine your 3D assets in Sequencer, include the final effects, and render out a high-quality movie scene
- Light your 3D movie set using Lumen lighting in UE5

Packt is searching for authors like you

If you're interested in becoming an author for Packt, please visit `authors.packtpub.com` and apply today. We have worked with thousands of developers and tech professionals, just like you, to help them share their insight with the global tech community. You can make a general application, apply for a specific hot topic that we are recruiting an author for, or submit your own idea.